CONTENTS

FOREWORD

When I was a kid, I was bored. My mom let me have only one hour of screen time a *week*—so there was a lot of time to fill. Sharon, Massachusetts, was small; we didn't have a single traffic signal in the entire town while I was growing up. I was an only child until my brother, Jonathan, came along years later, and then eventually my sister, Sara, the author of this book. So I would spend time running through the outdoor spaces, building forts, making "inventions" out of cardboard, and reading books. Being bored forced me to fill my own world with curiosity and creation, and this hands-on approach to science and technology served me well later in life when I started companies such as Siri (which made the voice assistant in Apple products) and Change.org, the world's largest social action petition website.

One of the things I was most curious about was how our bodies work. I decided that the human body had to be the most miraculous creation on the planet! When I was working

How Does My Body Work?

Human Body Book for Kids

HOW DOES MY BODY WORK?

HUMAN BODY BOOK *for* KIDS

STEAM Experiments & Activities
FOR KIDS 8-12

Sara LaFleur, MD

Illustrated by DGPH Studio
Photography by Nancy Cho

Z KIDS • NEW YORK

*This book it dedicated to all children—
who are the best thing about my life and who
fill me with hope and curiosity every day.*

*Special love and gratitude for Zoe and Isla, two
incredible humans who make parenting look easy.*

Copyright © 2021 by Penguin Random House LLC

All rights reserved.
Published in the United States by Z Kids, an imprint of Zeitgeist™,
a division of Penguin Random House LLC, New York.
penguinrandomhouse.com

Zeitgeist™ is a trademark of Penguin Random House LLC

ISBN: 9780593196946
Ebook ISBN: 9780593435434

Illustrated by DGPH studio (Diego Vaisberg & Martin Lowenstein)
Photography by Nancy Cho
Book design by Katy Brown

Printed in the United States of America

1 3 5 7 9 10 8 6 4 2

First Edition

on Siri, I realized how hard it was to build a machine that could listen to sounds, convert sounds into words, and then understand the meaning behind the spoken request; this is something almost every three-year-old can do without even thinking! Working on robots made me recognize how far scientists are from being able to build something even close to what the human body can do. Do you realize that, unlike robots, we don't need to be plugged into a power source because we contain our own portable energy generator (mine was powered by peanut butter sandwiches)? And if we bruise or break something, an entire army of cells gets to work on protecting us from infection and self-healing the injured area. I have no idea how to make a robot do that.

I would have loved to have a book like this one during my childhood. It is filled with weird information and fun Science, Technology, Engineering, Art, and Math (STEAM) activities and experiments to help you better understand the miracle machine that is your body. Take the surprising journey through each subsystem; you will be glad you did!

Adam J. Cheyer

INTRODUCTION

Did you know that you are amazing? It's true! (And not just because your parents say you are.) You are the superstar owner and boss of one of the most incredible creations ever made—your body! Imagine *trillions* of little unseen parts inside of you working in sync to make you *move, think, hear, see, taste, breathe, stay healthy,* and *grow.*

Welcome to your very own guidebook that will help you puzzle out what is happening and how things work in your body. When I was a fifth grader, I was already dreaming about being a doctor: being brave in emergencies, solving mysteries (making medical diagnoses), and having people trust me to take care of them. Now that I have grown to become Dr. Sara LaFleur, I can't wait to share with you all of the interesting things I have learned along the way.

You may have already learned some basics about the human body and its systems in school. Some things may have seemed a little scary, kind of icky, or just unbelievably cool. Either way, I am so excited that you have decided to learn even more with me and have fun doing it. Together we will do a bunch of interesting Science, Technology, Engineering, Art, and Math (STEAM) activities and experiments designed just for kids your age that will make it easy to understand the important roles body systems play in keeping us alive. Don't worry! You will be able to find most of the tools and materials you need at home to do the activities in this book.

Get ready to meet the human body's entire cast of superheroes (such as the heart and lungs), explore the battle between good (the immune system) and evil (viruses and some bacteria), and adventure into the weird questions you have always wanted to ask (like, *why do I turn red when I'm embarrassed?*). Turn to the next page and let's go!

BUILDING BLOCKS

As we begin our adventure together to uncover the mysteries of how the human body works, try to recall a time when something incredible happened with the help of a team. Take the launch of SpaceX into orbit as an example. Did you know that more than 8,000 scientists came together, each with their own specific contribution, to enable that safe, smooth travel into outer space? Your body works in much the same way. Trillions of individual cells inside you gather like building blocks to form special organized teams called tissues and organs. They work in specific jobs called body systems to accomplish an impressive goal: keeping you alive.

CELLS
Small But Mighty

Your body is made of structures called cells that are different shapes and sizes depending on their function in the body. Using a microscope, you could see that muscle cells are diamond-shaped and can squeeze and relax to move your bones around. On the other hand, red blood cells are Frisbee-shaped to glide along delivering oxygen and nutrients. Each human cell is made mostly of water, has the same important things inside, and is covered by a protective layer called a membrane.

Think of the cell plasma membrane (the cell cover) as the crossing guard at school. The membrane directs who may enter the cell (energy and food) and what things are ready to exit (waste products). The nucleus is like the school principal; it supervises all cell activities. The nucleus also holds the most important instruction manual of the cell, the chromosomes and DNA, with all the specific information that makes you unique. The mitochondria are like the school chef, cooking up energy for the cell and working alongside the Golgi apparatus (a coiled-up processing center) that packs up the food and the endoplasmic reticulum (a transport

Cells are organized into tissues that make up organs.

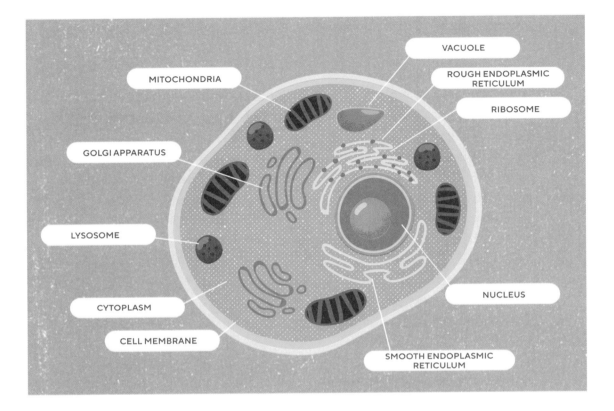

system) that delivers the packages to the lunch line. The lysosome (filled with chemicals that can break down molecules) can be thought of as your friend who eats any of your leftovers and the ribosome (makes proteins) as your friend who brings extra lunch meat to share. The vacuole (stores cell waste) acts like the compost and recycle bins where you place your trash after your meal.

All these cell components are housed in a substance called cytoplasm that looks like the lemon gelatin you finished for dessert. But how do all those tiny specks come together to make a person?

TISSUES
Like-Minded Buddies

Groups of cells in your body get packed up into organized units called tissues. Think about a time when you combined small Lego pieces of the same color or size (like cells) into several layers (like tissues) and stacked them to build a final masterpiece to play with.

There are four types of tissues that can be found in the human body. First, a type of tissue that glues everything together is called connective tissue. Bones, blood, fat, and cartilage are connective tissues. Next there is the epithelial tissue that provides a covering, such as skin and the lining of your mouth. Third, muscle tissues add movement to the body. Muscles can be either the type that acts on its own, called involuntary smooth muscle (the heart beating), or the type called striated voluntary muscle that you can control with your brain (the biceps muscle to lift something). Finally, there is nervous tissue (your brain and spinal cord) that carries messages and communicates with all the different parts of the body. The human body is an amazingly well-designed bundle. What goes inside the package?

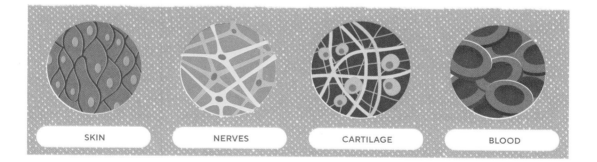

SKIN   NERVES   CARTILAGE   BLOOD

ORGANS
The Body's Shining Stars

Underneath the protective and supportive layers, you will find the body's superheroes: the organs. These guys are made from the different types of tissues and perform many of the mind-boggling feats in the body such as keeping you safe from toxins (liver) and preparing the air you breathe (lungs). Organs are so important in the human body that five of them (brain, heart, kidneys, liver, and lungs) are considered vital, meaning you could not stay alive without them.

Some organs in the body work in pairs while others stand alone. All the organs work in teams called body systems to get jobs done. Apart from feeling your heart beating after running or hearing your stomach growling when you are hungry, you may not think too much about all the work the organs are doing, without stopping, behind the scenes. In one day, your heart muscle squeezes about 100,000 times, your lungs take about 20,000 breaths, and your kidneys clean more than 200 quarts of blood. Talk about superpowers! All this happens seamlessly throughout your lifetime while you play, study, work, or sleep.

WHOA, WEIRD!

Doctors can use tissues from animals, such as pigs and cows, to repair injuries such as burns and torn ligaments in humans. This is called a xenograft.

When your body grows, the cells aren't actually getting bigger. Instead, they multiply in number by dividing into two until there are billions from just one single starter cell.

BODY SYSTEMS
Team Players

In order to straighten out all the complex tasks required in the human body, we group similar organs that cooperate into body systems. If you have ever worked on a group project at school, you know that the most important first step is organizing which task each student will do before turning in the final assignment. Think of body systems this way too, working individually but then together to complete an overall goal of keeping you healthy.

Each chapter of this book gives you a bird's-eye view into these incredible systems that make up who you are. You'll learn what each system does, which major organs are involved, and how the parts work together to make a living person. You'll also discover some weird and wonderful facts along the way. And perhaps best of all, you'll find fun activities and experiments in each chapter that illustrate the concepts you just learned.

WHOA, WEIRD!

You are about half an inch taller in the morning than in the evening! Over the day, the connective tissues in your spine and leg joints get squished down by gravity.

Not all cells in the body are alive. After a cell dies, it displays an "eat me" signal on its surface and is gobbled up by a blob-shaped cell called a phagocyte.

FLEX YOUR TISSUES

Stretching exercises are a great way to keep muscle and cartilage tissue healthy, but why? Use this before-and-after science experiment to explore the benefits of warming up your muscles.

TOOLS & MATERIALS

› rubber band
› measuring tape
› freezer
› rectangular cardboard box with top flaps (about 2 feet wide x 2 feet high x 4 feet long)
› roll of masking or packing tape
› pencil and paper to record measurements
› exercise or yoga mat (optional)

THE STEPS

1. Test the flexibility of a warm "muscle" versus a cold one (rubber band): Stretch a rubber band between your fingers and notice its elasticity and flexibility. Measure how far you can stretch the band without it snapping. Now put the rubber band on a flat shelf in the freezer for 12 hours. Take it out and stretch it between your fingers again. What do you notice?

2. Make a sit-and-reach box: Tape three of the top flaps of the cardboard box shut. Place it up against the wall with the remaining flap pointing toward you.

STANDING
FORWARD FOLD

BUTTERFLY

RUNNER'S
LUNGE

FIGURE 4

What were the results of your experiment comparing warmed up muscles versus stiff or cold muscles? Flexibility is when your muscles and joint tissues can move easily through a wide range of positions without stiffness. When you have good flexibility, you are less likely to injure your tissues and are able to keep good posture. Stretching exercises help to warm up muscles by increasing body temperature and blood flow to the cells and preventing tearing or snapping (just like what happened to the cold rubber band). After stretching, you can do many more physical activities safely.

3. With your hands pressed together, lean forward as far as you can. Can you touch the side of the box? If so, have a family member mark where you touched it. If not, have them make a mark where you touched the flap.

4. Now perform the standing forward fold, butterfly, runner's lunge, and figure 4 stretch for 30 seconds each. Repeat each stretch twice for a total of three repetitions.

5. Perform the sit-and-reach test again. Measure and record how far your hands could reach this time. What did you observe?

WHAT'S FOR DESSERT?

Surprise your family after dinner and show off your cell biology knowledge by creating these fun gelatin cellular 3D models.

TOOLS & MATERIALS

› clear glass bowl
› 6-oz box of light-colored gelatin mix (cytoplasm)
› note cards
› pen or pencil
› 4–6 raisins or similar-sized candies (mitochondria)
› sprinkles (ribosomes)
› 4–6 candy leather strips (endoplasmic reticulum)
› small bag of mini M&Ms or mini round gummies (vacuoles and lysosomes)
› 4–6 smile-shaped gummies (Golgi bodies)
› large, round piece of candy such as a large gumball (nucleus)

THE STEPS

1. Use your glass bowl to represent the cell membrane. Prepare the gelatin according to the box instructions and cool it in the refrigerator for 15 to 20 minutes to firm up.

2. Make note cards listing the cell components and their functions while you wait for the gelatin to firm up to a soft, thick mix.

3. Take the bowl out of the refrigerator and add your candies carefully to the gelatin. Compare the bowl with the cell diagram on page 12.

4. Place the bowl back into the refrigerator for a few hours to solidify.

5. After dinner, present the dessert to your family. Use your note cards to teach them about what you find in a human cell before you split the gelatin up into portions. See if you can explain the functions of the plasma membrane, nucleus, mitochondria, vacuoles, Golgi apparatus, endoplasmic reticulum, ribosomes, and cytoplasm to your family while you gobble your treats together.

HOW AND WHY

Great job constructing your 3D art cell! You're more likely to remember what you have learned by building a model and teaching someone about it. Plus, it's more exciting than memorizing facts, especially if you get to eat your finished product for dessert.

PICK YOUR FAVORITE ORGAN AND PIXELATE IT!

In this activity you will play around with the tech and art focus of STEAM. Pick your favorite organ (heart, lungs, brain, or kidney) or one that seems cool that you'd like to learn about in greater detail. You will create your own pixelated graphic of the organ you selected.

TOOLS & MATERIALS

> computer
> printed picture of your favorite organ
> sheet of graph paper
> pencil with eraser
> colored markers

THE STEPS

1. Choose an organ to use for your pixelated art project.

2. Print a simple picture of the organ. *Tip:* search for "kids' image of the brain" (or whatever organ you chose).

3. Place your printed picture next to your graph paper as a reference.

4. On your graph paper, use a pencil to put a mark in the squares that will be colored in to match your printed image. For the mark, pick the first letter of your color—for instance, for a blue square, mark 'B'.

5. Leave empty squares for areas where you don't want color, including borders or lines between colors.

6. Using your colored markers, fill in each square with the color you marked it to be.

HOW AND WHY

More and more, you are using computers and technology as part of your schoolwork and creative projects. In this activity, you chose a human body organ that you are excited to learn about and created your own pixelated image of it, in the same way your computer does. To form an image on the screen, your computer reads a mark (either a 1 or a 0) in a grid form. If a mark is present, the computer will light up that square (called a pixel). If a zero is present, it will keep the light off. Millions of pixels (computer lights) come together to make an image you can see. Which organ did you pick to focus on, and what would you like to learn about it? Read on in the book to see if you can find your answer!

EGG-CELLENT CELL MEMBRANE EXPERIMENT

Cells in the human body are too small to be seen without a microscope, but that doesn't mean you have to miss out on the fun of playing with a cell that you *can* see. Take an up-close look at a chicken egg cell membrane to discover how osmosis (movement of water across a cell membrane from a higher concentration to a lower one) works in this simple science and math experiment.

TOOLS & MATERIALS

› flexible tape measure (sewing tape)
› 4 eggs
› pen and paper to record measurements
› 4 widemouthed glass jars or plastic cups (Make sure the egg fits through the opening easily with plenty of space to spare.)
› bottle of white vinegar
› food coloring (Neon colors are fun, or choose some colors of the closest holiday, such as orange and black for Halloween.)
› plate

THE STEPS

1. Measure around the center of each egg and record the diameter. Place one egg in each of the containers. Add vinegar to each container until the eggs are submerged; it's all right if they float up. Add several drops of food coloring to each container (one color per container). Let the eggs sit for 24 hours covered in liquid.

2. Pour out the foam and the liquid. Pour new vinegar into each container, making sure the egg is covered completely. Let sit another 24 hours.

3. On the third day, your eggs are ready to examine. Gently remove the eggs from their containers and place them on a plate. Note that the hard shell has dissolved, leaving a soft membrane covering the egg cell.

4. You will also see that the colored water has moved from surrounding the egg cell into the cell itself, making it change color and grow larger. What is the diameter of the egg now?

5. Break one of the egg cell membranes (or look at the inside of one that broke by accident). Can you see the cell nucleus in the egg yolk, a little pale spot called the blastodisc?

6. Wash your hands and work area with warm water and soap when you are finished handling the raw eggs.

HOW AND WHY

Why does the eggshell dissolve in vinegar? The vinegar is an acid that interacts with the calcium base of the eggshell to make carbon dioxide (that's why you saw fizzy bubbles form) and water. What is left is the soft cell membrane, which allows water to enter through a process called osmosis but blocks other particles. That's why you saw the egg swell in size and change color when the food coloring moved into the cell.

IS BIGGER REALLY BETTER?

In this science lab activity, you will puzzle out the reason that your cells are so tiny. Hint: It has to do with diffusion (movement of particles in and out of a cell). Make three different-sized cubes of gelatin, and watch diffusion in action. Which size cube allows for the quickest diffusion of particles?

Caution: Ask an adult for help using the ammonia (a toxic household chemical), the stove, and the kitchen knife.

TOOLS & MATERIALS

› raw beet or small red/purple cabbage
› kitchen knife
› pot
› 4 cups water
› oven mitts
› strainer
› 2 plastic or glass containers
› 1 teaspoon ammonia
› 4 packets of gelatin
› spoon
› cutting board
› ruler
› 2 cups vinegar

THE STEPS

1. Ask an adult to help you cut the beet or cabbage into quarters and place it in a pot with four cups of water to boil.

2. Turn off the heat when the water starts to boil and, with the help of an adult (and wearing oven mitts), pour the mixture through a strainer and collect the liquid in your glass or plastic container. Be careful not to burn yourself.

3. Add one teaspoon of ammonia to the liquid.

4. Add four packets of gelatin and stir the mixture.

5. Allow the mixture to set in the refrigerator six hours or overnight.

6. Remove the jiggly solid mixture from the refrigerator and turn the container upside down onto a cutting board. Using your kitchen knife, cut three different-sized cubes: 1 x 1 inch (regular), 0.5 x 0.5 inch (small), and 0.25 x 0.25 inch (tiny).

7. Pour two cups of vinegar into a container. Place the three cubes in the vinegar and observe how long it takes for the vinegar to diffuse into each cube and change it from a dark violet color to a lighter purple.

8. Which size cube changed color the fastest?

HOW AND WHY

In this chemical science activity, you observed that acid and base particles are attracted to each other. When you placed gelatin cubes made with a liquid chemical base (ammonia) into a solution of liquid chemical acid (vinegar), the acid and base molecules wanted to move toward each other. Building-block cells in your body are incredibly busy transporting all of the energy, nutrition, water, mineral, and waste particles across their cell membrane. Movement of these molecules in and out of cells (diffusion) allows all of your body systems to function properly and cooperate together as one big system. In the case of cells, bigger is not better. Molecules move in and out of tiny cells faster and easier than large cells.

THE NERVOUS SYSTEM

The nervous system is your window into the world and all of your experiences. This body system sets humans apart from all other organisms. It enables you to think, create, communicate, feel, sense, and respond as no other animals can. Check out how your nervous system keeps you informed, inside and out.

NERVOUS SYSTEM FUNCTION
How You Think and Feel

The nervous system is all about sharing important messages from your environment with the command center of the body: your brain. From there, the brain makes decisions about what to do and directs your body to jump into action.

Picture the nervous system like one big game of telephone that you play with your friends, passing along news until it reaches the last person. Let's see how this works. When you step on a nail in your backyard, your foot feels the sharp sensation. Nerves in your leg carry that information to the spinal cord in your spine. The spinal cord brings the signal to the brain to be processed. Your brain decides that the nail hurts and sends directions back down to your muscles to quickly step off the nail. The nerves in your foot listen and complete the instruction loop by moving out of the way.

In addition to making quick decisions, your brain is also responsible for storing experiences as short- and long-term memories. This enables you to learn from past mistakes and successes and get smarter as time goes on.

Your five senses of sight, hearing, touch, smell, and taste are also part of the nervous system. And the nervous system manages your body temperature and balance.

Because the brain performs all of this incredibly important work, it needs daily time to rest and recover through sleep, dreams, and mindfulness practices such as meditation.

WHOA, WEIRD!

Dreams are a way for your brain to "clean up its room." Dreams allow your brain to clear unnecessary thoughts and keep important ones organized and stored in your long-term memory.

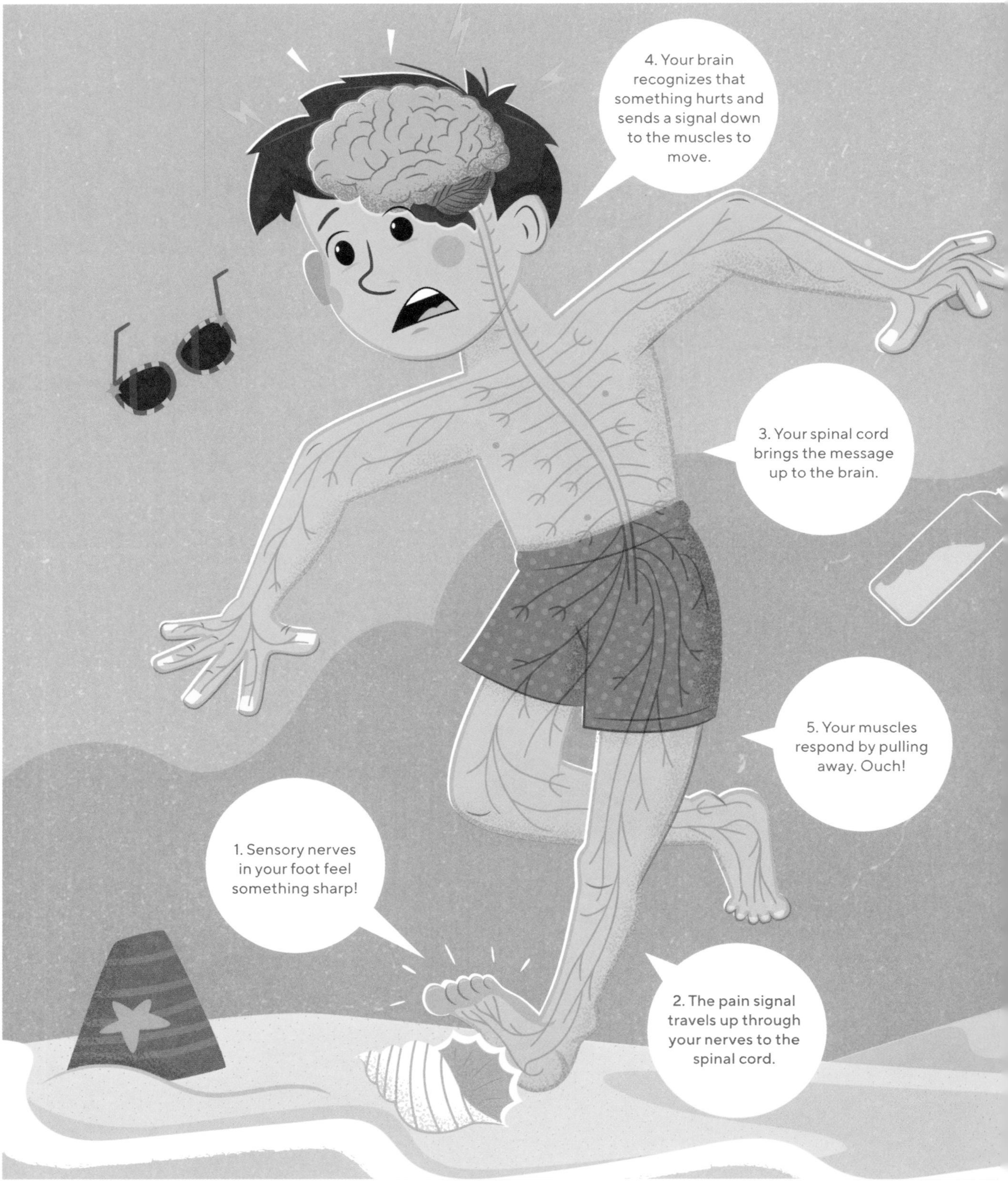
4. Your brain recognizes that something hurts and sends a signal down to the muscles to move.
3. Your spinal cord brings the message up to the brain.
5. Your muscles respond by pulling away. Ouch!
1. Sensory nerves in your foot feel something sharp!
2. The pain signal travels up through your nerves to the spinal cord.

CENTRAL NERVOUS SYSTEM
The Mastermind System

Just like in the cell, where the nucleus is in charge of what happens, the nervous system has a pair of structures that run the show. The brain and the spinal cord together are considered the central nervous system and are the main decision makers.

These two organs are vital to basic body function and are almost totally enclosed in bony protective layers. The brain and the spinal cord are soft and squishy like gelatin (which is why your skateboard helmet is so important!) and are surrounded by a special liquid called cerebrospinal fluid that helps to cushion them. This high-tech fluid also blocks anything from entering the brain and spinal column space that doesn't belong. Cells of the nervous system are called neurons, and groups of them make structures called nerves.

Let's take a closer look at the brain and the spinal cord.

BRAIN: THE CONTROL CENTER

What is it about the human brain that makes us unique? Scientists believe that our brains are wired unlike any other animal: information is passed among billions of nerve cells in our brain with electric bursts that allow humans to perform super-complicated tasks.

There are three main parts of the brain. Each has its own job to do, but they also work together to direct all of the body systems.

WHOA, WEIRD!

Did you know that mice brains are similar in structure to human brains? Scientists use mice to test medicines for treating human brain diseases.

The largest part of the brain is called the cerebrum and is divided into a left and right side of coiled brain tissue. The left brain controls the right side of the body and vice versa. This means that when there is an injury to the left side of a person's brain, they may not be able to move their right arm or leg. All of the tasks that you are consciously aware of—for example, thinking, sensing, feeling emotions,

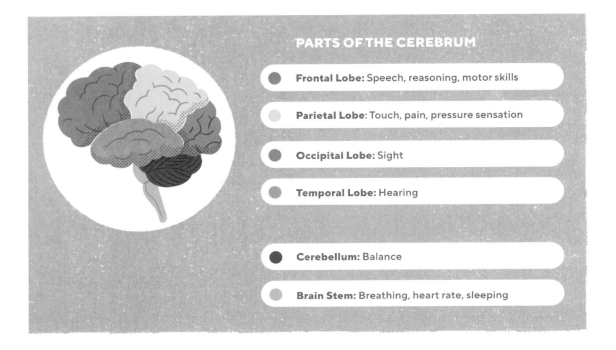

PARTS OF THE CEREBRUM

- **Frontal Lobe:** Speech, reasoning, motor skills
- **Parietal Lobe:** Touch, pain, pressure sensation
- **Occipital Lobe:** Sight
- **Temporal Lobe:** Hearing
- **Cerebellum:** Balance
- **Brain Stem:** Breathing, heart rate, sleeping

speaking, and moving your muscles voluntarily—happen in the cerebrum. When you are playing an instrument or studying facts for your science test, you are using your cerebrum. Memories are also stored there.

Underneath the cerebrum is a small ball of a spaghetti-looking structure called the cerebellum. Actions that you are unaware of happen here such as balance, walking, body position, and coordination (making smooth movements).

The cerebrum and cerebellum sit on a stalk called the brain stem, which connects and communicates with the spinal cord. The brain stem oversees body temperature, thirst and hunger, and heartbeat and breathing. It also reports news about pain, cold, and heat to the spinal cord.

SPINAL CORD: THE COMMUNICATION HIGHWAY

The spinal cord looks like a telephone cord or cable wire and acts like one too. It runs down from the brain through the circular bones in the spine to your lower back. The cord is made up of nervous system cells called neurons, and it carries information from the brain to the body and from the body back to the brain.

If you run your hand down the middle of a family member's back, you can feel the bony bumpy edges of the spine. Each hard spot that you can feel is another level of the spinal cord. At each of these spots from the neck to the tail bone, skinny string-shaped fibers called nerves run from the spinal cord to different areas in the body such as your arms and legs.

Sometimes there isn't time for a signal to travel to the brain and back down to protect you. When you need to pull your hand off a hot pan or close your eyes to block sand at the beach, a response called a reflex kicks in. In these instances, the information bypasses the brain and goes directly to the spinal cord and back to the muscles to move them. Treat your spinal cord with care as it's the only part of the nervous system that cannot repair or heal itself after an injury.

WHOA, WEIRD!

Some people are born without the ability to feel pain. Although this may sound good to you, pain is protective. A stomachache lets you know when something is wrong and motivates you to get help from your doctor.

Did you ever notice that you can't stop sneezing or blinking even if you try to? You can't control these reflexes. They are built into your nervous system to keep you safe from things that could irritate your nose or eyes.

PERIPHERAL NERVOUS SYSTEM
The Feeling System

Apart from the brain and the spinal cord, all of the other tissues in the nervous system are a part of a group called the peripheral nervous system. As you learned, cells of the nervous system are called neurons, and bundles of them make nerves. Peripheral means "on the edge of something." If you look at a map of the nerves in your body, you can see that you have nerves forming networks that branch like spiderwebs to cover every last inch of you! The peripheral nervous system is like a two-way highway with traffic traveling to and from the central nervous system.

Sensory nerves can feel pressure, pain, and temperature as you touch items throughout your day. They bring this information to the brain to process. Motor nerves carry orders back down from the brain to the muscles you want to move (somatic system) and to the organs inside your body that move without you thinking about it, such as the heart (autonomic system).

NERVES: A RELAY RACE

You might be wondering how nerves "talk" to each other. Each nerve cell has tree-like branching "feeler" ends called dendrites that can receive electrical impulses from a neighboring nerve cell. The electrical impulse, called an action potential, is made when salts move in

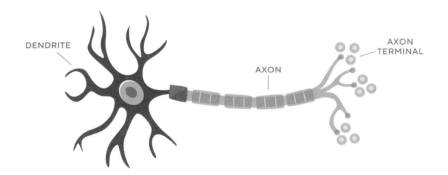

DENDRITE

AXON

AXON TERMINAL

and out of cells. This impulse travels down a thin cable called an axon. When the signal gets to the end of an axon, called the axon terminal, a chemical liquid called neurotransmitter is squeezed out. The liquid flows across the small gap between nerves and sticks to the next nerve cell, triggering another electrical burst. This relay race of transferring information from one nerve to the next continues around the body without stopping.

How does the electricity get inside your body to excite nerves? The movement of salts (that you eat) such as sodium, potassium, and chloride in your cells creates the electric energy that nerve impulses use. Light, sounds, pressure, and pain can also energize nerve cells. If you have ever sat too long on your foot and it went numb, you have experienced an upset in the normal nerve pathway. The nerves in your feet get squished and for a moment they cannot send signals to the brain. You will not feel anything until the nerve starts to recover, resulting in a pins-and-needles sensation.

SENSES: THE WINDOW TO YOUR WORLD

Your five senses—sight, hearing, touch, smell, and taste—also call the nervous system their home. Each of the senses detects a change in the environment (a stimulus) and sends the information to the brain to determine what is happening.

Smell and taste work together to sense chemicals released by foods or odors. The chemicals reach nerve endings in the tongue (taste buds) and the nose (olfactory receptors), resulting in different tastes and smells. As you may have noticed during your last cold when your nose was stuffed up, you might have had trouble tasting your food.

The tongue helps you identify different tastes.

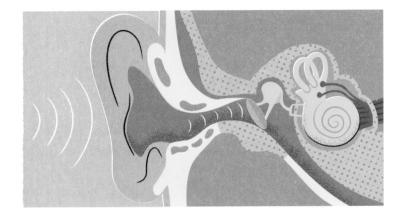

Hearing works by sensing sound waves (movement of air). The signal travels from the outer ear into the middle ear and bounces off the ear drum (a flat tissue inside the ear). Fluid in the inner ear then wiggles back and forth in a pattern that is understood by the ear's nerve endings. From there, the brain decides what you have heard.

Changes in light are sensed by the eye and allow you to see things. The eye is shaped like a sphere; the lens gathers light while the pupil adjusts how much light enters. Light-sensitive nerve cells are bunched together inside the eye in a structure called the retina, which sends information about images along the optic nerve to the brain.

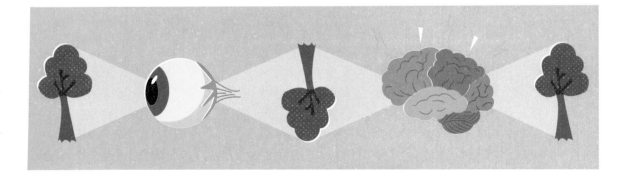

The sense of touch detects changes in pressure at nerve endings in the skin.

Finally, your sense of balance (often considered a sense) understands changes in body position. When you move, fluid sloshes around inside a spiral-shell-shaped space in your ear, and nerve endings there send a message to the brain. If the balance nerve endings get jostled too much, like on a rocky boat, you will feel motion sickness until they quiet down again.

Putting It All Together

Taking care of your nervous system starts with the foods you put in your body. You really are what you eat! Vitamin B, calcium, and potassium are important for transmitting nerve impulses along axons and can be found in fruits such as bananas, pomegranates, prunes, and oranges as well as in proteins such as meat, dairy, and eggs. Practicing ways to quiet your mind with mindfulness activities such as meditation or stretching is another way to boost your nervous system health.

WHOA, WEIRD!

The weird electric-shock feeling you get when you hit your "funny bone" is actually not from a bone at all. The ulnar nerve in your elbow runs very close to the surface of your arm and, when bumped, gives a startling pins-and-needles feeling.

EXPLORING YOUR SENSES: SIGHT

Through your senses, you get an understanding of everything around you. What would it be like to live without one of the five senses? Let's step into the shoes of a person living without sight. What other senses do you think may become even more important to someone who is blind?

TOOLS & MATERIALS

> blindfold
> piece of 8 x 10-inch cardboard or cardstock
> pencil with eraser
> braille alphabet chart (see page 37)
> 2 sheets of gemstone stickers (round shaped)
> hole puncher
> piece of string or rope (about 12 inches)

THE STEPS

BLINDFOLD RECOGNITION TEST

1. Gather several family members or friends in a room.

2. Ask someone to blindfold you.

3. Tell participants to line up in front of you, an arm's length away. Instruct everyone not to speak or give away any clues as to who is in front of you.

4. See if you can recognize who is who without your sight. Which senses did you rely on to guess?

MAKE A BRAILLE DOOR SIGN

1. Think of a short phrase like "Sara's Room" or a message that you would like to put on your door sign.

2. Turn your cardboard with the longer side lengthwise.

3. Lightly pencil in the letters for your sign (they will be erased later).

4. Using the braille alphabet chart, find the corresponding braille for each letter on your sign. Stick on the gemstones in the braille pattern for each letter until all letters are filled in.

5. Punch a hole in the top left and right corners of the sign. Thread your string through the holes and make a knot at each end to hang your sign.

6. Erase the penciled letters, leaving just the braille message.

7. Read your sign with your fingers, not your eyes.

HOW AND WHY

Although it's hard to imagine living without sight, hearing, touch, smell, and taste, this science activity shows how your brain can adapt and learn to use the senses that are available. For instance, you were able to use touch to recognize faces and read a door sign without seeing. Your brain is trained from experience to select the most important sense needed at the time.

A	B
C	D
E	F
G	H
I	J
K	L
M	N
O	P
Q	R
S	T
U	V
W	X
Y	Z

EXPLORING YOUR SENSES: TASTE

When it comes to trying foods, do you prefer salty items, have a sweet tooth, or wrinkle your nose at spinach? Are you sure? With this science experiment, let's try some tricks to fool your sense of taste. A lot more than you think goes into your ability to taste what you eat.

Caution: Ask an adult for help using the kitchen knife.

TOOLS & MATERIALS

› kitchen knife
› small potato
› small apple
› small onion
› plate
› small avocado
› blender
› tablespoon
› water
› 3 small bowls
› small banana
› ½ cup frozen spinach, thawed to room temperature
› saltine cracker
› graham cracker
› 3 spoons
› 2 paper towels

THE STEPS

1. Working with a partner, cut up two small cubes each of the potato, apple, and onion (skin removed). Leave these on a plate for later.

2. Remove the skin and pit from the avocado and put it in the blender with two tablespoons of water on high for one minute. Put this mixture in a small bowl. Repeat the same steps for the banana and spinach separately. Rinse the blender in between the different types of food.

3. Break the saltine and graham cracker in half.

4. Now you are ready to taste test with your partner! Hold your nose and close your eyes. Have your partner feed you the potato, apple, and onion cubes separately. Could you tell which was which? Hold your nose and close your eyes again and taste test the avocado, banana, and spinach purees on a spoon. Can you distinguish between them?

5. Take a sip of water and get ready for your last taste test. Wipe off your tongue with a paper towel. Make sure it is really dry. Have your partner feed you the graham cracker and saltine pieces separately. Be sure to wipe your tongue again with a paper towel between crackers. Did you know which was sweet and which was salty?

HOW AND WHY

Smell and sight work together to help the taste buds on your tongue taste foods. When you smell foods, chemical droplets in the air travel to your nose, and a signal about the type of food is sent to the brain. At the same time, visual hints about what the food looks like and what it has tasted like in the past are processed by the brain. Without these helper senses, you are fooled and may not know what you're eating!

The saliva on your tongue helps to dissolve the chemical particles of the food and allows you to recognize what it tastes like. When your tongue is dry, tasting becomes a challenge. It's no wonder you have difficulty tasting foods during a cold when your nose is blocked and your mouth is dried out!

READY, SET, ACTION (POTENTIAL)!

Neurons are where all of the action happens in the nervous system. These nerve cells send signals or messages called action potentials down a long rope (the axon) to the next cell. In this STEAM model, you will create your own neuron and watch it activate.

TOOLS & MATERIALS

› 10 feet of thin rope
› scissors
› 2 large plastic containers
› roll of toilet paper
› 4 table tennis balls or other lightweight bouncy balls

THE STEPS

1. Cut four one-foot pieces of rope. Tie a knot at one end of each piece. These will be your dendrites (the spindly feelers of the nerve cell).

2. Ask an adult to help you make a hole in the center of two plastic containers.

3. Thread all of the dendrite ropes through the hole in one container, threading from the outer bottom toward the inside of the container. Tie a knot inside the container to secure the ropes. This container will be your cell body.

4. Take your remaining six-foot-long rope and tie a knot at one end. Thread the unknotted end through your cell body container, threading from the inside of the container toward the outside. Tie a knot outside the plastic container. This rope will be your axon.

5. Load the roll of toilet paper onto the long rope by the cell body container. This will represent the action potential message that travels down the axon.

6. Add your second container to the other end of the long rope. Thread the rope into the bottom of the second container toward the inside. Tie a knot inside the container to secure the rope in place. This will be your end of the nerve cell (called a synaptic terminal).

7. Put the table tennis balls in the second container. These will be the cell neurotransmitters (chemical messages that are released to the next cell).

8. Ask one person to hold the cell body container and one person to hold the synaptic terminal container, open sides facing out. Stretch the rope tight. Have one or two others hold the dendrite ropes stretched out.

9. Let's start the nerve cell action potential! Throw a ball to the person holding a dendrite rope. When they catch it, they have received a chemical message from another nerve cell. Now it is time for the person who caught the ball to send the toilet paper roll (action potential) down the axon rope until it meets the end of the cell (synaptic terminal) and knocks the balls (neurotransmitters) out of the container. Those balls represent chemical messages that will travel to the next nerve cell (once you throw them) and start the cycle again. You have just seen how one nerve cell communicates with the next!

HOW AND WHY

Neurons are the basic cell units of the nervous system and allow for communication through-out the body. Building a super-sized version of this cell shows all the working parts including the dendrites, cell body, axon, synaptic terminal, and neurotransmitters up close. When a neuron is resting, no messages travel around the body. When a neuron is excited, an action potential travels down the axon and stimulates the release of neurotransmitters that "talk" to the next cell.

BUILD YOUR OWN THERMOMETER

The brain and the central nervous system are in charge of temperature control in your body. When you take your temperature with a thermometer, you are gathering important information about whether you are sick. If a fever (high temperature) is present, you probably are. Build a functioning thermometer here and check out how the science behind it works.

TOOLS & MATERIALS

› ruler
› black marker
› plastic or glass bottle
› water
› rubbing alcohol
› food coloring
› clear straw
› modeling clay
› large bowl
› pen and paper to record measurements

THE STEPS

1. Use a ruler to make ½-inch markings with a black marker on the side of your bottle from the bottom up.

2. Fill your bottle ¼ of the way up from the bottom with equal parts water and rubbing alcohol. Add several drops of food coloring to make the liquid easy to see.

3. Insert the straw into the bottle one inch from the bottom (in the water/rubbing alcohol mixture), and keep it in place by forming a seal around the straw with the modeling clay wrapped tightly around the bottle opening. An inch or two of the straw should stick out of the top. Keep the end of the straw open but try to make the seal around the bottle mouth airtight.

4. Dip the bottle in different temperature water baths (cold and hot) in the sink or a large bowl. Try warming the bottle up by holding your hands around its base. See how many inches the water rises or falls in the straw. Record your measurements.

HOW AND WHY

Thermometers work when heat expands the liquid (in this case water and alcohol), causing it to move up through a closed space (in this case a straw). When you have a fever, the liquid warms up, expands, and travels up the thermometer like on the model you built. If it goes past 98.6 degrees Fahrenheit, you have a fever.

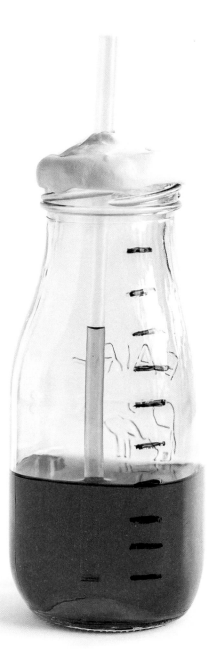

PUT ON YOUR THINKING CAP

Your brain is divided into sections that have different jobs to do. Here you'll create a wearable map of your brain that shows which area springs into action when you are skateboarding (balance), which one lights up when you are chatting on the phone (expressive language), and so on.

TOOLS & MATERIALS

› diagram of the brain sections (see page 30)
› black marker
› light-colored swim cap
› colored markers

THE STEPS

1. Using your brain diagram as a reference, draw sections of the brain with a black marker on the swim cap.

2. Color the sections and label them by name. Add the detail of the folded coils (gyri) that the brain is folded into.

3. Use a black marker to label the swim cap with some of the most important functions (such as language) that each part of the brain is in charge of.

4. After the marker is dry, put on the cap and model it for your family. Think about your usual weekday. Which sections of your brain are you using for the different activities you do?

HOW AND WHY

The frontal lobe, parietal lobe, temporal lobe, occipital lobe, and cerebellum work together to perform all the complex tasks required in your everyday activities. The left side of the brain controls the right-sided movements, and vice versa. This STEAM project combining science and art to make a mental map (your thinking cap!) helps you get a picture of what happens where in your central nervous system.

THE MUSCULOSKELETAL SYSTEM

Muscles (musculo–) and bones (–skeletal) join together in the musculoskeletal system to power all your activities throughout the day. Whether you are skateboarding, typing, dancing, playing piano, or smiling, large and small muscles are involved. Here is what you need to know about the body system that is behind all your awesome moves.

MUSCULOSKELETAL SYSTEM FUNCTION
Moving You Around

When you think about the musculoskeletal system, you might picture the movement and muscle strength of bodybuilders or athletes. However, this system also has some cool (and weird!) side jobs. Muscles attached to bones help you move, but you also have muscles deep inside the organs and blood vessels of your body. This hardworking crew has the less glamorous, but important job of propelling and squeezing stomach juice, poop, pee, and blood through your body.

Bones do more than just hold you upright against gravity and protect your organs from bumps and bruises. They produce your red and white blood cells and platelets (which we'll visit in the circulatory and immune system chapters). Bones also store the treasured mineral salts, calcium and phosphate, which your body needs to stay alive.

BONES
The Body's Piggy Bank

After seeing dried fossils in museums or skeletons on Halloween, you may not think that bones in our bodies are alive. But bones are organs that are constantly growing and changing and are filled with blood, nerves, and cells packed in a soft material called bone marrow. Marrow, called spongy bone, is the site of a special factory where new cells are made for the blood.

The hardened white outer layer of bone is called periosteum, or compact bone. There are all different types of bony shapes, from flat hip bones to seashell-shaped bones in the ear to long cylindrical

bones in the arms and legs. Bones must be adaptable in order to fit into a small space (when babies are inside the mother's uterus) and then grow to adult size. For this reason, some bones in the skull and pelvis start out as smaller, malleable pieces that fuse into one piece of harder bone later in life. Eventually, the length of bones determines our final height as grown-ups. Bone lengthening happens at a special pearl-colored area at the end of a bone, called the growth plate, where a shiny connective tissue called cartilage expands and turns into more bone over time.

Calcium and phosphate are salts that help keep our bones strong and hard. These mineral salts enter our body from the different types of foods we eat. Bones act like a piggy bank: they store calcium and phosphate when there is extra in the body and loans them out when needed (for example, to power the heart muscle).

WHOA, WEIRD!

Sometimes you need surgery for a broken bone and sometimes just a cast. Why? Ends of a broken bone that are far apart need to be pulled together by a surgeon with metal pins. Broken edges that are close to each other can heal themselves just by touching inside the protective shell of a cast.

Solve the mystery of the disappearing bones: A baby is born with 300 bones and has only 206 by the time they are an adult. Where did the extras go? Babies have to be flexible enough to fit through the birth canal. As babies grow, soft cartilage hardens, and the bones fuse together to make the body's final framework.

Do you know anyone who is double-jointed and can bend their fingers or thumb backward? In reality, no one has two (double) joints where there should be one. Instead, these people have inherited an increased stretchiness of the ligaments that allows the joint to move beyond its usual range.

MUSCLES
The Triggers

Your bones would be like a pile of pick-up sticks without your muscles to jump-start them into action. Muscles are made up of cells called myofibrils that are squished together to form rope-like bundles called fascicles. Each muscle has its own nerve, artery, and vein. When energy in the form of a chemical called ATP and glucose (sugar) is received from the blood, the muscle cells pull and squeeze tight, shortening the muscle length. Muscles work in pairs, pulling against each other to bend at the joint in opposite directions like a tug-of-war game. When the muscle relaxes, the joint straightens out again.

There are three main types of muscles. Skeletal muscles are attached to bone and can move your arms, legs, fingers, toes, neck, and other joints when your brain tells them to. Cardiac muscle, which works automatically, is one of the hardest-working muscles in the body, keeping your heart beating more than two billion times in an average lifetime. Smooth muscle lines the inside of your inner organs and blood vessels. This type of muscle also does its job automatically. The bladder, which tightens and relaxes repeatedly when you pee, is made of powerful smooth muscle.

WHOA, WEIRD!

The tongue is an unusual group of eight muscles wrapped together like a rope. It is the only set of muscles in the body that works independently from your skeleton and is not attached to bone.

CONNECTIVE TISSUES
Smooth Operators

Joints are areas where bending at two bones can occur. Connective tissues at these spots help hold the joints together and allow for smooth movements. Ligaments are like thick rubber bands that join two bones together. Tendons act like sturdy pieces of duct tape that hold muscles to the bone. Cartilage, as you read about earlier in relation to the growth plate, is part of the skeletal connective tissue and helps to cushion bones. Little gel pancakes of cartilage reduce friction and stress between the bones of your spine. You can also find cartilage at the joints to help bones glide against each other, as well as in the ribs, nose, and ears.

Putting It All Together

Keeping your musculoskeletal system healthy is easy and fun to do. This is your chance to jump around and burn off all your energy! The more active you are, the better condition your muscles and bones will be in. Exercise brings oxygen and important nutrients to your tissues. Try to devote one hour a day to your favorite way to get sweaty (soccer, dance, gardening, running after your dog, and so on). Don't forget to stretch afterward to prevent tearing or injury at the joints. Load up on calcium in your meals for strong bones and you will be all set to score goals in this body system.

WHAT MOVES YOU?

When you think of all the ways you can move, twist, and bend, you can understand that many different types of joints are needed in the body. In this activity, locate and observe several types of musculoskeletal joints in everyday objects. Then build a thumb joint.

TOOLS & MATERIALS

› salt or pepper grinder
› closet door
› swivel mop
› sliding window
› dresser drawer
› sheet of paper
› wooden craft stick or any type of stick
› tape

THE STEPS

Take a few minutes to see how the different items move by putting them into action. Compare them to your own body's movement:

1. Hold the salt or pepper grinder with two hands and twist the top back and forth as an example of how a pivot joint works, like the one that lets your neck turn side to side.

2. Open and shut a closet door to observe how a hinge joint works and rotates with a cylinder-shaped pin inside. Now move your forearm up and down to observe your elbow's hinge joint.

3. Look at the ball-and-socket mechanism of a swivel mop that lets you move the handle around in all directions, like this type of joint in your shoulder and hip. Try moving your arm around at the shoulder to see the range of motions it can make.

4. Slide a window up and down and notice how the hard surfaces easily glide over each other, like the gliding plane joint in your wrist.

5. Open up your dresser drawer and look at the fixed joint where the two pieces of wood come together at the front of the drawer. How are the wood surfaces interlocked to stay in place? This is the type of immovable joint that you find when two bony areas come together in the skull.

6. Build an example of an ellipsoidal joint, like the one in your thumb. Take a sheet of paper and fold the edge of it back and forth on itself to make an accordion fan. Tape the bottom of one side of the fan to a stick. Holding the stick steady, open the fan out to the left and right and close it back up and see how your thumb moves at the base of your hand.

HOW AND WHY

Although you aren't able to see inside the thumb, shoulder, hip, elbow, wrist, and skull joints in your body, this activity gives you an idea of how two hard bones meet and move smoothly with different mechanisms. Each type of joint enables a separate range of motion and set of activities.

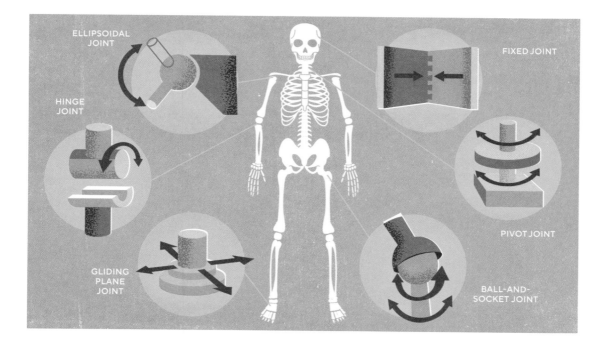

MUSCLE MANIA

Imagine you're one of the superstrong athletes on the Olympic weightlifting team. With this science-based art-and-coloring project focusing on musculoskeletal anatomy, you will create your own muscle suit to wear.

Caution: Ask for a parent's help choosing clothing to mark up for this project.

TOOLS & MATERIALS

› colored fabric markers
› plain, solid, light-colored long-sleeved T-shirt
› plain, solid, light-colored leggings, cotton pants, or pajama bottoms
› muscle group pairs reference sheet (see page 55)

THE STEPS

1. Gather your fabric markers and select as many bright colors as possible. Pick the same color for the right and left muscles of the same type (for example, red for both the right and left biceps).

2. Place your shirt and pants on a large flat surface.

3. Starting on the front of the shirt and pants, use different colors to draw the muscles from the chart on the following page. You can use a long oval shape for each muscle. Write the muscle name inside the oval.

4. When you have completed the muscles on the front of your shirt and pants, flip them both over and add the matching opposing muscle pairs on the back.

5. Try on your muscle suit to wear for the day. Check out which muscle pairs are helping you as you bend and straighten your body parts throughout the day!

MUSCLE GROUP PAIRS

Sternocleidomastoid (front) Bends neck forward	**Trapezius (back)** Straightens neck up
Biceps (front) Bends upper arm at elbow	**Triceps (back)** Straightens upper arm at elbow
Wrist Flexors (front) Bends wrist	**Wrist Extensors (back)** Straightens wrist
Deltoid (front) Lifts shoulder	**Latissimus Dorsi (back)** Relaxes shoulder
Rectus Abdominus (front) Bends at belly button	**Erector Spinae (back)** Straightens at belly button
Iliopsoas (front) Bends thigh	**Gluteus Maximus (back)** Straightens thigh
Quadriceps (front) Bends knee	**Hamstrings (back)** Straightens knee
Tibialis Anterior (front) Bends foot toward shin	**Gastrocnemius (back)** Straightens foot away from shin

HOW AND WHY

Muscles are organs that work in pairs to help you move at your joints (where two bones meet). In order for muscles to get strong enough for Olympic weightlifters to lift a record 500 pounds, the muscle pairs must do repetitive work (pushing or pulling) against resistance (a force that wants to slow something down). In this activity, you looked at the major muscle groups that work opposite each other (opposing muscles) to bend and straighten your bones. With your muscle suit on, you could see which muscles were helping you do all of the activities requiring strength throughout your day.

LEND A (ROBOTIC) HAND

Who doesn't need an extra hand to help around the house sometimes? In this STEAM activity you will build your own functioning hand that moves. To get this project going, start by looking at your own hand from the wrist to the fingertips, and count how many joints (places that move back and forth or can bend) you have.

TOOLS & MATERIALS

› pencil
› piece of cardstock or thin cardboard
› scissors
› 8 colored straws (one larger size if possible)
› hot or other liquid glue
› 5 pieces of different-colored yarn or embroidery floss, each about 1 foot long
› tape, if necessary

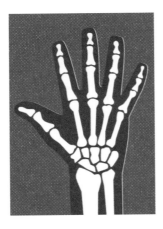

THE STEPS

1. Trace your hand with a pencil on your cardboard or cardstock.

2. Cut out the hand shape.

3. Make a mark at the wrist and finger joints (where two bones meet) with a pencil on the cardboard hand.

4. Bend the cardboard slightly at all the penciled joint markings.

5. Cut the straws to fit between the joints; they will act as the bones in your hand and fingers. You should end up with long bones in the hand, middle length bones at the bottoms of the fingers, and the shortest bones at the fingertips. Glue your straws in place on your cardboard and let dry. At the wrist area, cut and glue the larger straw (about two inches long) to hold all of the strings you will guide through.

6. Thread the colored strings through the straws, one color per finger. Tie a few thick knots at the end of the string toward the fingertip. If necessary, use

tape to prevent the string from slipping through the straw. Thread all of the strings through the larger straw at the wrist.

7. Pull on the strings to make each individual string move. Surprise your family or friends with your movable robot hand!

HOW AND WHY

With this engineering-based model, you can see tendons in action, moving fingers at the joint connections between two bones. For an advanced interactive model, you can complete the above activity on both sides of the cardstock to be fully anatomically correct. The flexor set of tendons that allow your fingers to bend run on the palm side of the hand, while the extensor set of tendons that straighten the hand run on the backside of the hand.

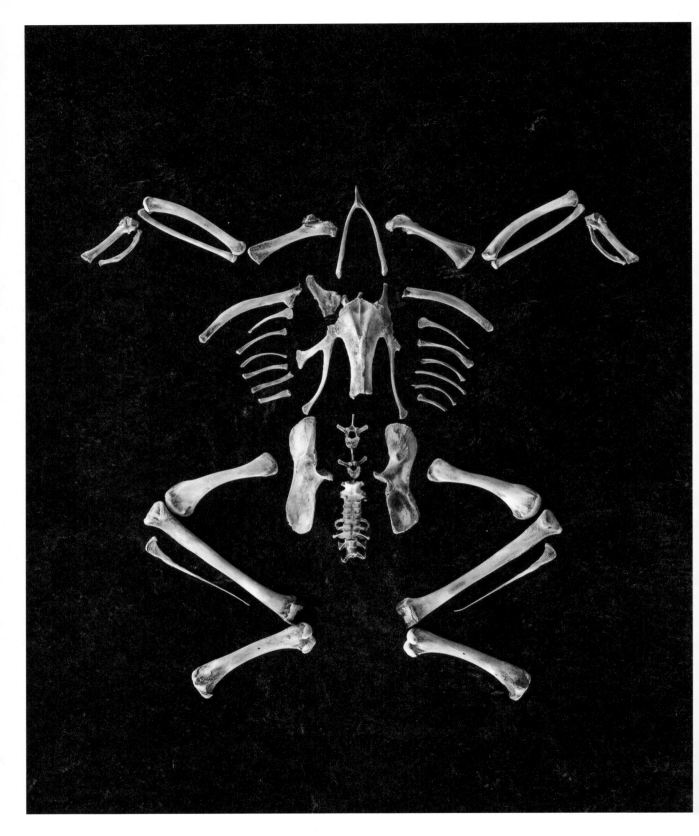

7. When you are done cutting the mask out you will have two layers to make a double thick mask. Remove the pins used for holding the layers in place. Try it on for size!

FACE COVERING EXPERIMENT

1. Gather a neck gaiter, a bandanna, and a fabric or paper face mask.

2. Ask an adult to help you light a candle, and put it in a holder or hold it in your hand a few feet away from your face.

3. With no face covering on, try to blow the candle out. Easy, right?

4. Put your neck gaiter over your mouth and nose and have an adult relight the candle. Try to blow the candle out. Can you still do it? What happens if you double up the gaiter fabric over your mouth and nose? What about with a bandanna? Can you blow out the candle when you wear the fabric or paper face mask?

5. Put on the face mask you made and try to blow the candle out. Which face covering makes blowing out the candle the hardest?

HOW AND WHY

Respiratory infections such as the common cold spread to other people in droplets from sneezing and coughing into the air. You are less likely to get someone else sick if you wear a face mask—rather than a neck gaiter—over your nose and mouth. Thick weave fabrics with multiple layers limit air with germs from traveling. They not only keep you from spreading an illness, but they also block any sneaky viruses or bacteria in the air from entering your upper respiratory tract, preventing you from getting sick.

GAS LAB

What exactly is gas when we talk about the gas exchange that happens in the respiratory system? Is there a way to see what carbon dioxide (the waste gas produced in cells) looks or acts like? Let's build a chemical science model to better understand gases.

TOOLS & MATERIALS

› funnel
› 2 cups white vinegar
› clear plastic bottle or container with label removed
› paper towel
› balloon
› 1 cup baking soda

THE STEPS

1. Using a funnel, pour the white vinegar into the bottle until it is about ⅓ full.

2. Dry the funnel with a paper towel.

3. Use the funnel to fill a balloon with baking soda until it is about halfway full.

4. Holding the balloon in your hand so no baking soda escapes, stretch the balloon end over the bottle opening, letting the top of the balloon droop over the edge of the bottle.

5. Stretch the balloon straight up so the baking soda slides down and mixes with the vinegar. Watch the chemical reaction as the balloon inflates with carbon dioxide gas that is formed!

HOW AND WHY

Oxygen enters your body when you breathe. Glucose (sugar) enters when you eat. These two molecules mix to make energy for your body that powers your cells. In turn, cells release water and carbon dioxide as waste, which you exhale. In this experiment, you mixed vinegar with baking soda to make carbon dioxide gas that blew up a balloon! Even though you aren't able to see carbon dioxide, you can understand its presence when it inflated a balloon.

TREASURE CHEST

Your ribs protect a pair of special treasures—your lungs! The lungs are vital organs, and without them, you could not stay alive. Every breath you take brings in precious oxygen that your cells need to function. Build a working lung model with this engineering experiment, and see what is happening inside your chest.

Caution: Ask an adult for help if you decide to use a box cutter.

TOOLS & MATERIALS

› plastic bottle with cap
› scissors or box cutter
› black marker
› 2 bendy straws
› glue
› tape
› 3 balloons
› rubber band

THE STEPS

1. Ask an adult to help you make a small hole with the tip of the scissors in the cap of a plastic bottle. The hole should be big enough to fit your straw.

2. Ask for help cutting off the bottom half of the bottle with either the scissors or a box cutter.

3. The top half of the bottle will be your chest. Draw ribs on the left and right side of the bottle with a black marker.

4. Stretch out the ribbed part of a bendy straw. Bend the straw end to the side. Make a small hole in the ribbed part of the straw.

5. Cut a two-inch piece off another straw. Insert this piece into the hole in the first bendy straw. This will make a Y shape. Trim the straws so that the Y arms are even. Add some glue around the second straw if needed to keep it in place.

6. Turn the Y upside down to look like the trachea and the left and right bronchus of the lungs. Tape one balloon to each of the straws that point down. Make sure the balloon–tape connection is snug around the straw so no air escapes.

7. Put the upside-down Y in the bottle and thread the straight straw through the bottle cap. Twist the cap down tight. Seal it with glue if necessary.

8. Tie the bottom of the third balloon in a knot. Cut a small piece of the top of the balloon off.

9. Stretch the balloon over the opening at the bottom of the bottle. If needed, use a rubber band to secure the balloon in place.

10. You are ready to begin breathing! Pull down on the knot of the balloon at the base of the bottle. This is your diaphragm muscle working to inflate the two lungs (balloons) in the chest. The straw through the bottle cap represents your main windpipe, the trachea, which branches into the two straws (the right and left bronchus).

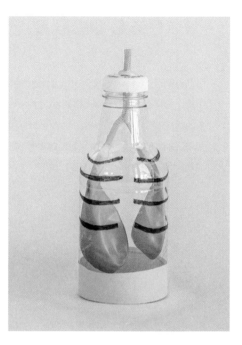

HOW AND WHY

You take about 960 breaths per hour. The diaphragm pulls down at the bottom of the chest, and the lungs spring open. Air travels down the trachea and into the left and right bronchus to the lung tissues, where oxygen exchange happens.

VOCAL CORD BANJO

The human voice is unique to each person. An individual can be identified just by the sound that travels through their vocal cords and voice box (larynx). What makes your voice sound the way it does? Create a vocal cord banjo with this science and art activity, and test out why some voices are thin and high pitched while others are deep and low.

TOOLS & MATERIALS

› 1 bottom of a gift or pencil box, about 1 x 3 x 8 inches
› 3 or 4 rubber bands of different widths
› pencil

THE STEPS

1. Stretch the rubber bands across the box lengthwise.

2. Slide a pencil under the rubber bands across the box.

3. Pluck the rubber bands in front of the pencil. Slide the pencil up and down the length of the box and strum in front of the pencil. How does the sound change if the rubber bands are short and tight versus loose and long?

4. Play your banjo for a family member or friend, and explain how vocal cords work with the following explanation.

X-RAY ART

In this activity, you'll blend the science and art aspects of STEAM. Human anatomy (the study of the body) can be inspirational for creating new works of art. A chest X-ray taken at the doctor's office can provide a window into how the lungs, ribs, and diaphragm look. It's also a beautiful piece of black-and-white art showing the smooth shapes of the respiratory system! Gather a few items from your kitchen and craft shelf to create your own X-ray art masterpiece using the image on the next page as a guide.

Caution: Ask an adult for help using the kitchen knife.

TOOLS & MATERIALS

› white art paint
› small painting or other type of tray
› sheet of black construction paper
› marshmallow
› handful of string beans or fava beans
› large potato
› kitchen knife
› dish or bath sponge
› paintbrush

THE STEPS

1. Pour some white paint into your tray. (Use the paint sparingly so that it doesn't drip on your paper. Use your food items the way you would use an art stamp.)

2. Set up your black construction paper with the long side from top to bottom (portrait).

3. Lightly dip the side of the marshmallow into the white paint. Starting at the bottom center of your paper, about one inch up from the bottom, press the marshmallow on the paper to make a square-shaped imprint. Travel up the paper, stamping the marshmallow in a straight line for a total of 12 squares (or however many fit). These will be your spine bones.

4. Gently dip one of the beans into the white paint. Press it onto the paper to make the shape of a rib connected to the top spine bone, angled slightly down. Keep adding ribs, one on the left and one on the right, attached to all of the spine squares on your page.

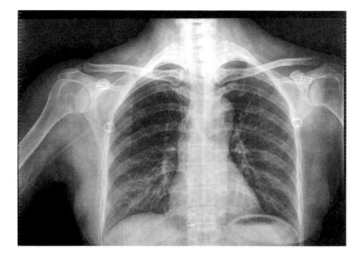

5. Dip two fatter beans into the paint and press them at the top of the page on the left and right to represent the collar bones.

6. Ask an adult to help you cut a potato lengthwise and crosswise into four pieces. Dip the potato into the white paint and press it at the bottom of the page with the curved side up to make an imprint for the diaphragm. Make one stamp on the right and one on the left. You should have a half-moon shape.

7. Gently dip another potato quarter into the paint and press it in the center of the page to make a stamp of the outline of the heart.

8. Wring out a damp sponge and dip it into the paint. Gently press it over your ribs on the left and right side of the paper to represent the lungs.

9. Use your paintbrush to outline the shape of the chest and fill in any additional details you like.

HOW AND WHY

X-rays—images that are formed when an energy beam is directed at an area of interest—are one of the main ways to determine if the respiratory system is healthy. Tissues that are dense and block the X-ray, like bone, appear white, while tissues that let the energy pass through easily, like the lungs, show up gray. X-rays are used to check for illnesses such as a pneumonia infection, which looks like a white fluffy cloud in the lungs. In your X-ray art, you can see the bones of the spine, the twelve ribs on the left and right, the two collar bones, the two spongy lungs, the diaphragm muscle, and the heart in the center of the chest.

THE DIGESTIVE SYSTEM

The digestive system has an important and pretty gross job: it takes the food you eat and breaks it down into a nutritious mush that your organs, tissues, and cells can use, and gets rid of the rest as poop! Here's the cool science behind how your next meal turns into energy for your body.

DIGESTIVE SYSTEM FUNCTION
Moving Food Along

The digestive system has three main chores to complete for the body: digestion, or breaking down foods into small particles; absorption, or taking up nutrients from food into the blood; and elimination, or discarding waste materials.

The word digestion means to break substances down into smaller parts. The body uses two main methods to break down foods. Mechanical (physical) breakdown starts when your teeth and tongue rough up and chop up foods in your mouth. From there, the muscular tube (the digestive tract) churns, moves, and tosses the food about until it turns into a nutrient-filled liquid that can nourish cells. It carries food through the esophagus, stomach, and intestines until it exits through your rectum. This squeezing movement of the smooth muscles is called peristalsis.

Chemical breakdown of food happens at many separate areas in the digestive tract, including in the mouth, stomach, gallbladder, pancreas, and intestines. The body produces several different types of really cool chemicals called enzymes that can help speed up the digestion of food through a series of steps called biochemical reactions. When enzymes wash over small pieces of food in the digestive tract, they disintegrate the particles into the smallest possible nutrients that cells can use.

Specialized folded surfaces called villi in the digestive system, which are covered by microscopic blood vessels, take these good particles from the digestive juices and pass them into the blood for transport.

After the food has been broken down and the vitamins, minerals, and other healthy molecules have been absorbed into the blood, a liquid mush

WHOA, WEIRD!

Your digestive system produces about three tall glasses of gas a day! Burps and farts are your body's way of making sure too much gas isn't trapped inside your body, causing pain. The stinky odor is sulfur produced by bacteria in your large intestine.

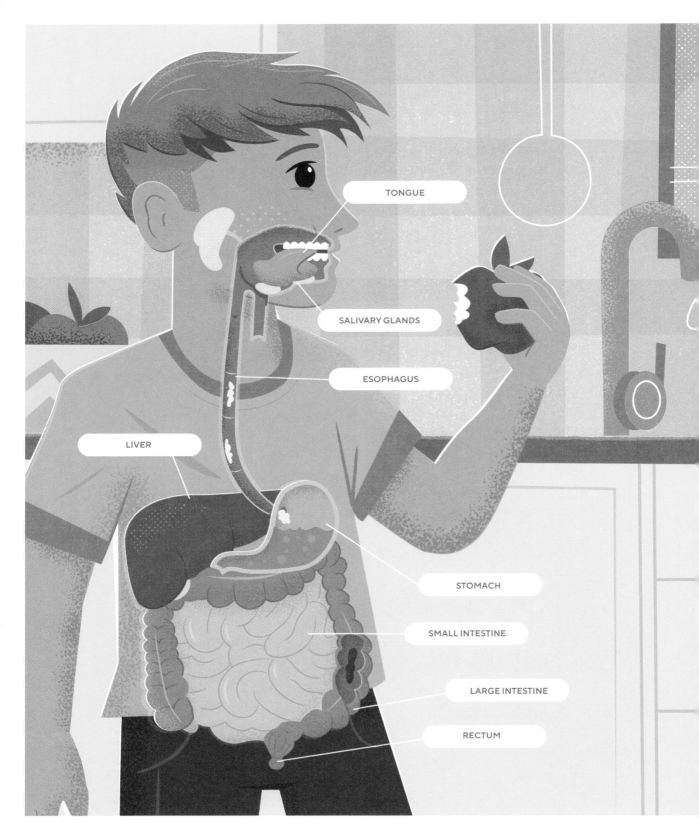

remains. The last job completed by the digestive system is to get rid of what is not needed by your body. Water is recycled and taken back up into and used by your cells, and a semisolid waste—your poop—is what remains. When this material exits your rectum through your anus, the digestive process is complete! All together, the digestive tract is one long continuous tube through your mouth, esophagus, stomach, small and large intestines, and rectum. Read on to see what happens as your meal travels along the way!

MOUTH
Power Player

You might think of the mouth as the first stop on the digestive system road trip, but the digestive process really begins when a reflex in your nervous system is triggered even before you take a bite. When you smell something delicious, an area in your brain called the medulla sends signals to the salivary glands under your tongue and behind your jaw to release saliva. Your mouth is watering, and you are now ready to eat!

In the mouth, several body parts work as a team to transform food into the liquid nutrition that cells can absorb. Your front incisor teeth rip and tear while your back molars grind food into bits. The tongue, a powerful group of eight muscles wrapped together, pushes food toward the back of the mouth before it enters the esophagus.

Salivary glands release liquid saliva that helps to moisten food, allowing chemicals from foods to interact with the tongue and be tasted. A dry mouth and tongue can make it hard to taste foods, as you may have noticed during your last cold. You can't see

WHOA, WEIRD!

Have you ever had food poisoning? It's not the food itself but rather harmful bacteria carried with the food that causes the illness. Your body knows to protect you by vomiting or causing diarrhea to quickly rid itself of the pesky intruder.

the salivary glands because they hide under the tongue and behind the jaw. They squirt saliva into the mouth through small openings called ducts.

ESOPHAGUS
A Slide to the Stomach

The esophagus is a long muscular tube that connects the mouth to the stomach. It has the important job of blocking strong stomach acid from backing up into your throat. At the bottom of the esophagus is a tight ring of muscles called a sphincter that opens when food is ready to travel through it and closes when the stomach is churning. The esophagus also helps to coat food with a slippery liquid called mucus for easy movement.

Your body has a built-in protective mechanism called the epiglottis that prevents food from going down the airway instead of down the esophagus on its way to the stomach. The epiglottis is a flap of tissue that closes over the opening to the airway when food passes by. The involuntary muscles of the esophagus squeeze and relax to move food along until it reaches the stomach.

STOMACH
Your Personal Food Processor

Your stomach is tucked under your ribs on the left side of the chest (not behind your belly button). It is a flexible muscular organ that can shrink to the size of a tennis ball when empty or stretch to the size of a J-shaped football after a big meal.

WHOA, WEIRD!

Every five to seven days, the stomach and intestines make an entirely new lining to repair themselves from all the wear and tear of mechanical and chemical food processing.

The liquid produced by your stomach cells, called hydrochloric acid, is so strong it can dissolve a piece of metal. Its purpose is to disinfect and break down the food you eat.

Special cells that line the inside of the stomach produce a very strong liquid called hydrochloric acid. This acid not only breaks down all types of foods you eat, but also sterilizes foods, removing any harmful bacteria that may have traveled along with your food. The acid in the stomach stays put with two valves that close shut, protecting both your esophagus and your small intestine from the harsh chemicals that could irritate their tissue lining. The stomach uses both mechanical (churning and squeezing) and chemical (strong acid) methods of processing foods.

Sounds from your stomach, such as gurgling, growling, and sloshing, are a normal part of the digestive process. Your stomach has to work extra hard to mush up foods that you have not chewed well into small pieces. When pressure builds in the stomach from air that enters during swallowing, the stomach releases the gas as a burp to avoid a stomachache. All this noisy commotion tells you that your body is busy at work!

SMALL INTESTINE
Deceptively Big

The small intestine is a muscular, tube-shaped organ that is divided into three sections. The sections have different jobs and funny-sounding names: duodenum, jejunum, and ileum. The intestine is coiled up like a bowl of spaghetti in your abdominal (belly) cavity. But the small intestine is not actually so small. In total, the small intestinal tube can be up to 30 feet long in adults! It's called small because it's skinny compared to the thicker, wider large intestine.

Liquid from the stomach enters the duodenum, or first part of the small intestine. Here more chemical digestion takes place. Organs, such as the gallbladder and the pancreas, squirt substances into the duodenum through little tunnels called ducts that help break down fats, larger carbohydrates, and proteins.

In the jejunum and the ileum, digestive juices come into contact with millions of little folds of intestinal lining called villi (as described earlier), which absorb the nutrients into the blood.

The small intestine is considered a vital organ, meaning you could not live without it. When you make healthy food choices, your small intestine happily gobbles up all of the vitamins, minerals, proteins, and other essential energy molecules your body needs to run all of its systems.

LARGE INTESTINE/RECTUM
Processing Poop

The last stop that your meal takes through your digestive system is the large intestine (also called the colon). Any materials that could not be digested are stored here. Water, some useful salts, and any minerals that have not yet been saved by the small intestine are absorbed in the large intestine.

The large intestine has its own microbiome, or community of 100 trillion good bacteria that live in it. Without the help of these guys, you would not be able to make vitamin K, which prevents bruises and bleeding in the body. Additionally, these bacteria help to further process your poop. When your body is ready to get rid of the waste it can't use, the solid material moves down to the last section of the large intestine, the rectum. You can sense the pressure in your rectum; this is your signal that it is time to use the bathroom. Poop exits through a ring of muscles called the anus.

LIVER/PANCREAS/GALLBLADDER
The Digestive Dream Team

A team of three more organs pitch in to do their part to process food as well as store nutrients as energy for times when the body is fasting (not eating). Each digestive organ has its own set of enzymes that it adds to the soupy digestive mix to speed the chemical breakdown of a meal.

THE LIVER: BIG AND MIGHTY

Your liver is the largest organ in your body and is a powerhouse of digestive activity. Nutrient-carrying blood from the small intestine travels to the liver to be sorted out. Any harmful substances that were produced during the breakdown of foods and medications are

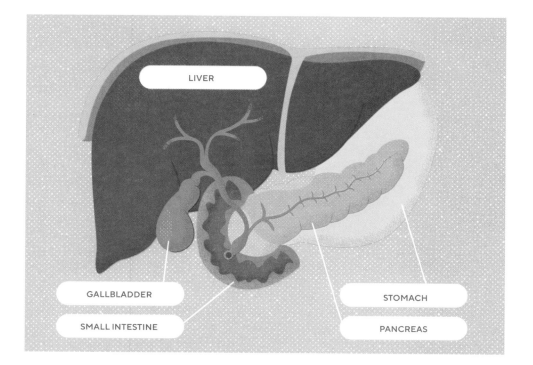

LIVER

GALLBLADDER

SMALL INTESTINE

STOMACH

PANCREAS

cleaned out by the liver. Extra energy molecules are stored in the liver and can be released into the blood whenever the body needs them. The liver also recycles old blood cells and makes bile acids for fat digestion.

THE PANCREAS: SUGAR PATROL

The pancreas works hand in hand with the liver to keep a delicate balance of the sugar level in your body. Some sugars (which come from carbohydrates such as pasta and bread) are necessary to power your cells. However, too much sugar in the blood can cause a disease called diabetes which, over time, can poison other organs.

When sugar levels start to rise after a meal, your pancreas releases a chemical called insulin into the blood. Insulin helps to transport sugar into cells that need it and safely stores the rest in the liver as glycogen for use later. The pancreas also has its own recipe of enzymes that it squirts into the small intestine to aid in digestion.

THE GALLBLADDER: LEAN AND GREEN

The gallbladder is a small, green pouch that hides under the liver. The main function of the gallbladder is to store bile acids that the liver makes until you are ready to eat a fatty meal. When you bite into a hamburger, your gallbladder is triggered to release a liquid into the small intestine that surrounds and digests the extra fat droplets that will be passing through. While the gallbladder is helpful, it is not essential. In some people, the gallbladder becomes clogged and needs to be removed.

WHOA, WEIRD!

Are you carrying any rocks inside your body? Some people form little hardened stones in their gallbladder based on the types of foods they eat. These usually don't cause trouble unless they get stuck in the little passageway into the small intestine, causing a blockage.

Putting It All Together

After reading this chapter, it should be clear that you truly are what you eat. Packaged foods with added chemical ingredients make your digestive system work harder to find healthy nutrients, vitamins, and minerals to feed your cells. Eating fresh fruits and vegetables as well as simple plant and animal proteins, like beans and meat, make it easy for your body to stay healthy and thrive. Likewise, drinking water (rather than sugary beverages such as juice or soda) helps move foods along in your digestive tract and prevents gas and waste from backing up and causing a stomachache. Practice cooking with your family at home so you can think about and see what is going into your meals. Finally, make sure that raw proteins such as eggs and meats are well cooked to prevent harmful bacteria from entering your digestive system.

WHAT'S GOING ON IN THERE?

A lot of what we now know about stomach digestion comes from scientific observations made by William Beaumont, an army surgeon who lived during the early 1800s. One of his patients had a wound in the stomach area that allowed him to see directly how the stomach processed foods. In this science activity, you will use a less icky method to check out why chewing and acid are so important for the stomach's ability to break down what you eat.

TOOLS & MATERIALS

› 4 clear glass jars or cups
› masking tape
› pen
› 4 colored hard candies
› plastic baggie
› hammer or mallet to break up the candies
› 1 cup water
› 1 cup white vinegar

THE STEPS

1. Set out four clear glass jars. Label two of the jars Water and two of the jars Acid.

2. Place a whole candy in one of the jars labeled Water and a whole candy in one of the jars labeled Acid.

3. Using a plastic baggie and a mallet, crush up one of the hard candies. Repeat with another hard candy.

4. Place one crushed candy in the jar labeled Water and one crushed candy in the jar labeled Acid.

5. Pour ½ cup of water into each of the jars labeled Water.

6. Pour ½ cup of vinegar into each of the jars labeled Acid.

7. Observe which of the cups dissolves the candy the fastest. Which one took the longest?

HOW AND WHY

Digesting food into smaller usable particles for use by the cells in your body requires both mechanical (chewing, squishing, and crushing) breakdown by your teeth, tongue, and stomach as well as chemical (reactions between two substances) breakdown by the hydrochloric acid that your stomach cells produce. Vinegar is almost as acidic as stomach acid and will break down the hard candy more quickly than the water. The crushed hard candy will disappear faster than the bigger solid candy because smaller particles allow a larger surface area to be in contact with the liquid that will dissolve it.

CHEMICAL REACTION WIZARDRY

Your body performs all sorts of chemical wizardry and creates enzyme potions in order to extract the nutrients and healthy substances it needs from your last meal. With this STEAM activity, show a friend or family member the magic of how bile salts from the gallbladder help to digest fats.

TOOLS & MATERIALS

› 2 cotton swabs
› concentrated dish soap
› paper towel
› circular plate or pan with edges
› 2–3 cups whole milk
› food coloring, several different colors

THE STEPS

1. Roll one cotton swab lightly in a few drops of the dish soap on a paper towel.

2. Ask your partner (a family member or friend) to pour milk onto a circular plate or into a pan until it contains about ¼ inch of milk.

3. Give your partner a clean cotton swab, and take the soap-dipped cotton swab for yourself.

4. Add one drop of each of your food coloring colors to the milk in the pan.

5. Have your partner gently dip their cotton swab in the center of the milk without mixing it around. (Nothing should happen.)

6. Place your soap-dipped cotton swab in the center of the milk for several seconds and watch the color burst! Was your partner surprised?

7. Explain the magic of your chemical reaction to your partner with the following explanation.

HOW AND WHY

The digestive system uses different chemical reactions to help process the foods you eat. Bile acids, which are made in the liver and stored in the gallbladder, cause fat globules to break into microscopic droplets. In this activity, you wowed your partner when you added soap to the fat droplets in milk. The soap acted like a bile acid and dispersed the colored drops into a colorful burst. The cotton swab with no soap did not react with the colored milk drops.

WHAT GOES IN MUST COME OUT

The digestive (or gastrointestinal) tract is a 30-foot-long tube that starts at your mouth, coils up inside of your abdomen, and ends at your anus. It takes a meal about two to five days to travel all the way through your body. Nutritious elements and water are absorbed, and waste exits as poop. Let's make a digestive model to replay the journey your after-school snack takes.

TOOLS & MATERIALS

> a few favorite snack crackers
> banana
> resealable plastic bag
> ½ cup water
> ½ cup orange juice
> scissors
> 1 leg of a pair of tights or nylon stockings (cut from a pair)
> bowl
> paper cup
> pencil

THE STEPS

1. Place a few handfuls of your favorite snack crackers and a banana in a resealable plastic bag, and add ½ cup of water (mouth saliva).

2. Add ½ cup of orange juice (enzymes and stomach acid) to the resealable plastic bag. Press out any extra air and seal the top of the bag.

3. Mush and massage the food around with the liquid. This is how your stomach first processes and breaks down your snack.

4. Let the snack sit for about 30 minutes in the bag (stomach).

5. Cut a small hole in the corner of the resealable plastic bag. This will be the pyloric sphincter, the hole that foods travel out of and into the small intestine.

6. Squeeze the contents of the resealable plastic bag into the leg of your stocking. Place the stocking over a small bowl. Tie a knot at the top of the stocking so both ends are closed. The stocking works like the small intestine to hold your snack and absorb the water and nutrients from it.

7. Squeeze your small intestine stocking until all of the liquid is removed into the bowl. Set these "nutrients" aside. The rest of the solid mush that is left will move to the large intestine.

8. Poke a circular hole through the bottom of the paper cup with a pencil. This will be your large intestine, ready to receive the solid waste from your small intestine stocking.

9. Snip the end of the stocking and squeeze the contents into the paper cup.

10. Push the solids through the small hole (the rectum) at the end of the cup. This step models the last task of the digestive process—making poop!

HOW AND WHY

In this activity, you modeled the path your snack takes from start to finish through the digestive tract. In the mouth, saliva is mixed with enzymes to start initial breakdown of sugars. Next, the stomach churns the food with acids like a washing machine to make it easy for the small intestine to absorb nutrients and energy molecules into the blood. After all the healthy liquid is absorbed, a semisolid mush remains. The large intestine saves the water from these remains and moves the waste through as poop, finishing the cycle. With this engineering STEAM project, you created your own chemical and mechanical digestion model, showing the processing of two different foods, just like your body would.

SPAGHETTI SQUASH

Measured in a straight line, the tube that forms your gastrointestinal tract is about as tall as three basketball hoops stacked on top of each other. It's pretty incredible that all thirty feet of your esophagus, small and large intestines, and rectum fit inside your body like a squished-up string of spaghetti! Trace the path of the digestive tract in this STEAM art project, and design how you think the tube is arranged to keep it from tying in a knot or twisting on itself.

TOOLS & MATERIALS

> large piece of butcher paper (the length of you, from your head to your toes)
> colored markers
> 5 lengths of different-colored pieces of yarn (1 foot, 4 feet, 16 feet, 8 feet, 1 foot)
> tape measure or ruler
> scissors
> glue

THE STEPS

1. Ask a partner to trace the outline of your body with your head turned to the side on a piece of butcher paper.

2. Measure and cut your five different-colored pieces of yarn to the lengths in the Tools & Materials list. You will make one long connected string (through the next several steps) using the five different-colored pieces. Your diagram will follow the path of food through the digestive system.

3. Glue one of the one-foot pieces of string to your paper to run from the mouth to the bottom of your neck area. This will be the esophagus.

4. Connect the four-foot piece of string to the first piece of string, and then glue it to your paper in the chest area to represent the stomach. You can shape this "stomach" string into a J shape like it would look in your body.

5. Connect the 16-foot piece of string to the previous piece of string, and then glue it onto your paper to fit in the abdomen area as your small intestine. What shape do you think will keep it from tangling on itself?

6. Connect the eight-foot piece of string with glue as your large intestine, and glue it to the paper. It also needs to fit in the abdominal (belly) area. Where would you place it to stay out of the way of the small intestine?

7. Glue the last one-foot colored string to the previous string at the lower belly area. This piece represents the rectum, or last segment of your digestive tract.

8. Fill in any details that you like on your body tracing (such as eyes, hair, and nose) and label the different string pieces: esophagus, stomach, small intestine, large intestine, and rectum.

HOW AND WHY

The intestines are designed so that the total surface area (with lots of folds and microfolds) is huge! With this type of structure, the digestive liquids come into contact with many digestive system tissues and have several opportunities for nutrients to be absorbed. This way, no vitamins, minerals, or energy molecules from your meal are wasted. This is the reason our intestines are designed to be so long and folded up in the first place. The small intestine is arranged like an accordion and is attached to the abdominal wall so it won't twist. The large intestine surrounds the small intestine in an upside-down U shape to make room in the cramped belly space.

FOOD DETECTIVE

Chemical breakdown of food by substances called enzymes happens at different places in the digestive system. Each place in the digestive system has its own enzymes released there until all the different foods (carbohydrates, proteins, fats) that pass by are processed. The salivary glands in your mouth release the first important enzyme for digestion. Using the chart on the next page, can you do some detective work in this simple science activity to puzzle out which enzyme it is?

TOOLS & MATERIALS

› small cube of deli meat
› stopwatch or phone timer
› small cube of cheese
› small saltine cracker

THE STEPS

1. Review the chart on the next page that shows the action of three digestive enzymes.

2. Place the deli meat in your mouth and wait 30 seconds.

3. Observe any changes (taste or texture) in the food. If you take the food out of your mouth after 30 seconds, is it the same or different as when you put it in?

4. Repeat with the cheese and then the cracker.

5. Did any of the foods change shape, texture, or taste when they came into contact with saliva and the enzyme found in your mouth?

6. Based on your observation and the following chart, can you guess which enzyme is released by the salivary glands?

DIGESTIVE ENZYMES

amylase	breaks down carbohydrates to sugar
lipase	breaks down fats in cheese
pepsin	breaks down protein/meat

HOW AND WHY

Neither the piece of cheese nor the meat significantly changed or began to be broken down in the mouth. (Pepsin produced in the stomach breaks down proteins such as meat. Lipase from the pancreas and stomach help to digest fats in cheese.) But amylase produced by the salivary glands in the mouth reacted with the carbohydrates in the cracker, breaking it down to simple sugars. You may have even tasted a sweetness in your mouth when the cracker was digested by amylase.

THE EXCRETORY SYSTEM

Imagine the mountains of trash in a landfill. Your body has no space to store so much waste and must find creative ways to keep what it needs and dispose of the rest. Together, the urinary (kidney) and integumentary (skin) systems make up the excretory system, whose purpose is to get rid of liquid waste from your body. In other words, this is how and why you pee and sweat.

EXCRETORY SYSTEM FUNCTION
Getting It All Out

The excretory system describes the group of organs—the kidneys and the skin—that eliminate (get rid of) liquid wastes that are not needed in the body.

The kidneys' role is to filter and clean the blood. Just the way water filters clean tap water, the special cells in your two kidneys trap wastes and extra salts that your body doesn't need. As blood moves through the kidneys, wastes are sorted out and discarded in the urine (your pee!) that the kidneys create.

Acting like a coat of armor, your skin keeps out harmful bacteria, dirt, and other toxins (unhealthy substances) found in your everyday environment. Your skin is a living organ that can adapt and change to situations by opening up its pores (small holes in the tissue) to let out moisture and heat or squeezing tight to keep you warm. When you sweat, skin lets go of any salts your body no longer needs.

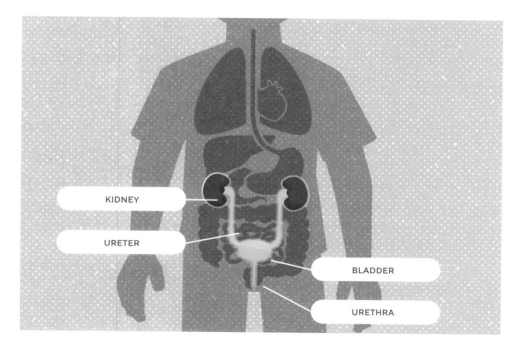

KIDNEY

URETER

BLADDER

URETHRA

In addition to removing waste, the excretory system keeps the environment inside the body the same, even though conditions on the outside of the body may be changing. This process is called maintaining homeostasis. For instance, in order for your cells and organs to function properly, they must be at a very specific temperature for chemical reactions to take place smoothly. The excretory system makes sure that no matter what the outdoor temperature is, your body's inside temperature stays steady. With homeostasis, the excretory system also balances acid-base, fluid, and mineral levels in your blood.

KIDNEYS
The Body's Filter

Tucked under your ribs on either side of your spine are two bean-shaped organs called the kidneys. They are two vital organs that you cannot live without for long periods of time. Their job of filtering and cleaning the blood is critical; without your kidneys, extra fluid and waste from your cells can build up and become poisonous to your other organs.

The kidneys are remarkably efficient at conserving water and nutrients. All the blood in your body travels through the kidneys several times a day. In total, your kidneys filter and clean about 600 cups of blood a day but discard only four cups of waste as urine.

The kidney is made up of millions of individual specialized cell units called nephrons. Each nephron has

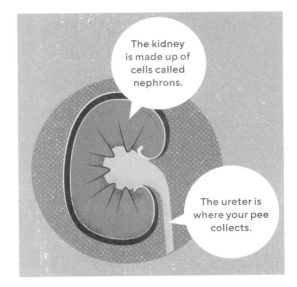

The kidney is made up of cells called nephrons.

The ureter is where your pee collects.

a coiled set of tubes that blood is squeezed through. They trap water and healthy minerals in the body and discard a small amount of waste—extra water, salts, and nitrogen—as urine. On their own, water, salt, and nitrogen are not harmful, but problems can result when too much accumulates inside your organ systems. Doctors can test what is in your pee and look at it under a microscope to determine if your body systems are functioning properly.

The kidney directs homeostasis, the natural balance in your body, as described above. When the kidney senses the blood pressure is too high or too low, for example, it adjusts the amount of water flowing in the body to make a "just-right" amount.

Your kidneys also help make red blood cells by sending a chemical signal (hormone) to the bone marrow to produce them. They also help strengthen your bones by absorbing minerals and making vitamin D.

URETERS
Slides for Waste

The two ureters are long skinny tubes that run from the center of your left and right kidneys down to the base of the bladder. When urine is formed, it collects into a funnel-shaped area in the center of the kidney called the renal pelvis and then drains down to the ureters. Occasionally, a stone or collection of minerals forms in the kidney, rolls down into the ureter, and gets stuck. Ouch! When a stone gets stuck in the ureter, urine can't travel past it and pressure backs up in the kidney, causing pain. Sometimes the stone can pass on its own, and other times a doctor needs to remove the stone.

BLADDER AND URETHRA
Holding Pee Until Go Time

The bladder is a hollow muscular organ shaped like a balloon. The bladder receives urine from the two straws attached to your kidneys (the ureters). The bladder can stretch to hold up to eight cups of pee during the night and about four cups before you have the urge to use the bathroom during the day. Once the bladder is fully stretched, it sends a signal along the nervous system to the brain that it is time to pee.

The urethra connects to the bladder and is the tube through which you pee. Although the bladder muscle is not voluntary and you can't control when it squeezes to empty itself, you *can* control when pee exits the body. At the end of the urethra, a small ring of muscles called the external sphincter holds your pee in until you decide you are ready to use the bathroom.

SKIN AND SWEAT GLANDS
Hot and Cold Gatekeepers

On a hot sunny day or when you play sports, you may notice that you're sweating quite a bit! You create heat in your body when your muscles and organs are working hard. Just like the kidneys' role in keeping the environment inside your body balanced, the skin has a role in homeostasis. When the temperature rises inside your body, the skin responds by opening up blood vessels close to its surface and letting heat escape. Sweat formed by the skin releases water to cool down the surface

WHOA, WEIRD!

Have you ever been embarrassed and turned bright red? Signals from your brain tell your blood vessels in the skin on your face and neck to open very wide (dilate) and blood rushes in. Scientists still don't know why (but they do know how) you blush!

Unlike humans, dogs have very few sweat glands (just a few in their paws). In order to cool down, they have to pant heavily. Moisture is lost from their tongues instead of their skin.

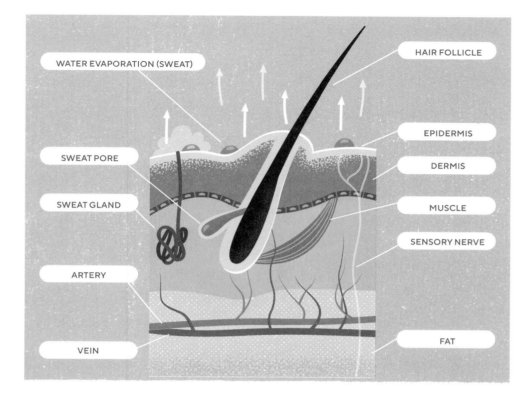

of your skin. Alternatively, on a cold day, the skin squeezes its blood vessels tight and holds onto heat.

The skin is your largest organ and partners with the kidneys to get rid of extra water, salts, and nitrogen (in the form of ammonia). The skin has three major layers: the outer epidermis, the deeper middle dermis, and the bottom layer of fat cells.

The deepest layer of skin is filled with fat cells and is designed as insulation to keep you warm. Sweat glands, hair follicles, nerves, and blood vessels run through this layer. Blood vessels bring cell waste, mixed with water in the sweat glands, to the surface of the skin to exit the body as liquid perspiration. The outer epidermis layer of skin is composed of dead cells that form a protective tough barrier against bacteria, dirt, and other toxins in the environment.

Three chemicals that are different colors mix together to give skin its color. Varying amounts of melanin carry a brown pigment, carotene carries a yellow-tan pigment, and hemoglobin carries a pink pigment.

Fingernails and toenails are a part of your skin and are made of a clear, hardened material called keratin. Keratin cells are dead and do not have nerve endings, which is why it doesn't hurt to cut your nails.

Putting It All Together

Two thirds of your body is made up of water. This means that one of the best ways to keep the environment inside your body balanced is by drinking enough water (two liters per day). Staying hydrated helps your excretory system flush out wastes through the kidneys. Your skin takes a lot of wear and tear every day, and if its surface barrier breaks down (with cuts or rashes), bacteria can enter. A sunburn can also injure the top layer of cells. Wearing sunscreen and moisturizer helps protect skin cells from damage. Also, keeping up with regular showers and cleaning your skin prevents sweat glands from getting blocked and developing an irritation such as acne or a rash.

BE COOL

Sweating is one of your body's strategies to cool off when its temperature is rising. Sweating transfers energy in the form of heat from your body to the air through a process called evaporation. Does a liquid that evaporates quickly make you feel cooler than one that takes more time? Let's find out in this science activity.

TOOLS & MATERIALS

› several cotton balls
› ¼ cup water
› ¼ cup rubbing alcohol

THE STEPS

1. Find a volunteer for your experiment.

2. Tell your volunteer to close their eyes and extend their arms, with wrists facing up.

3. Dip one cotton ball in water and one cotton ball in alcohol. Gently swipe one cotton ball across each of your volunteer's wrists.

4. Ask your volunteer which wrist felt cooler to the touch. Did the alcohol or water feel cooler?

HOW AND WHY

Sweating helps regulate your body temperature. When you sweat, water from your sweat glands is released to the air through evaporation. This process changes water from a liquid to a gas and requires energy in the form of heat. This energy is lost to the environment, and you feel cooler. Liquids such as rubbing alcohol, whose chemical property allows them to evaporate faster, make your skin feel cooler than water because the heat transfers away from your body more quickly.

NO WASTE CHALLENGE

Your body considers the chemical element nitrogen a waste, and the excretory system helps gets rid of it in the urine. In humans, nitrogen wastes can form ammonia that damages brain cells. But your pee can help take care of Earth! Did you know that you can recycle and repurpose your pee to make a superfood for your garden plants? While nitrogen compounds can be toxic in the human body, they can actually be beneficial to plants.

TOOLS & MATERIALS

> 2 pots
> soil
> packet of cucumber or tomato seeds (spring/summer) or cabbage seeds (fall/winter)
> glass or plastic container
> 2 spray bottles
> masking tape and marker to label bottles and pots
> index card
> ruler

THE STEPS

1. Fill both pots with soil, leaving ½–1 inch space from the top.

2. Open your seed packet, and plant half in one pot and half in the other based on the packet directions for depth in the soil.

3. Pee into a container that your parent says is acceptable to use. Wash your hands.

4. Ask your parent to help you add pee to one of your spray bottles along with water to make a 1:10 mix (1 part pee and 9 parts water). Discard any extra pee and wash your hands.

5. Fill the other spray bottle with water only. Label the spray bottles: one with Pee, one with Water Only.

6. Place both of your plants in a sunny spot. Label one pot A and one pot B.

7. Water pot A daily with the spray bottle that has pee in it. Water pot B daily with the spray bottle that just has water in it.

8. Make a note on your index card about which plant grows higher and is more robust (more leaves and growth). What did you discover?

HOW AND WHY

Human urine contains high levels of nitrogen, potassium, and phosphate, which are vital nutrients for some types of plants. Studies have shown that plant growth can be boosted by the minerals found in urine. In this science activity you performed a zero waste environmental challenge and found a useful way to repurpose something that would normally be considered waste: your pee!

ENGINEER A KIDNEY FILTER SYSTEM

Some people inherit a medical problem that causes their kidneys to fail. A kidney transplant can save their life. Until the person is able to receive a new kidney, they must have a way to clean liquid waste from their body with the help of a blood-filtering machine called dialysis. Use your STEAM skills to engineer your own version of a dialysis machine that can filter a pail of dirty water into a clean one.

TOOLS & MATERIALS

> newspaper or towels
> 3 scoops of dirt
> pail of water
> stick
> 4 glasses
> large funnel
> scratch paper
> pencil
> small measuring cup

The following are suggested materials. Feel free to choose your own supplies as part of your filter design.

> coffee filter
> cotton balls or gauze pads
> plastic or metal screen mesh or a small wire colander
> sand
> small pebbles
> steel wool pad or fibrous sponge

THE STEPS

1. Spread out newspaper or towels over your work surface to catch any mess while working.

2. Mix several scoops of dirt into a pail of water with a stick.

3. Place a large funnel into a glass to receive clean filtered water. Set aside three glasses to use later.

4. Brainstorm on scratch paper how to design a filter to clean your water from muddy to clear. What factors are important to consider?

5. Engineer up to four different filter systems to compare. Set the filter system up so that a cup of water travels through your system into the funnel and collects in the first glass. Repeat with the funnel and the remaining glasses, using a different filter system each time.

6. Which materials are the most durable? Which provide the cleanest water? Does the length of time the water spends trapped in the filter affect how clean it becomes? Which size of filter material lets muddy water travel through but catches contaminant (dirt) particles?

HOW AND WHY

Dialysis machines are special filter systems that help clean blood for people whose kidneys have failed. In this activity you explored the important factors to consider in the design of your filter. The opening or pore size in your mesh is ideally around ⅛ inch to trap particles. Thicker or layered materials help to increase the time a liquid is in contact with the filter, resulting in cleaner water. Some materials can easily tear when liquid or particles travel through and are not durable for larger amounts of liquid to be filtered. What was your most successful design?

SHOW ME YOUR TRUE COLORS

You know that the kidneys are in charge of keeping acids and bases balanced in your body, but what exactly are those? In this activity, you will learn about chemistry by making your own indicator paper that will change color to blue when touching a base and red when touching an acid.

Caution: Ask an adult for help with the kitchen knife and hot water on the stove.

TOOLS & MATERIALS

› head of red cabbage
› kitchen knife
› pot of water
› slotted spoon
› 5 coffee filters
› several paper towels
› scissors
› a few drops each of apple juice, lemon juice, cola/soda, pickle brine juice
› a few drops each of milk, dish soap, toothpaste, baking soda mixed with 1 teaspoon water

THE STEPS

1. Carefully chop up a red cabbage head and place it in a pot of water.

2. Simmer for 30 minutes.

3. Remove the pot from the heat, take out the cabbage with a slotted spoon, and discard it.

4. When the broth has cooled, soak your coffee filters in it for a few seconds. Then place them on a couple of layered paper towels to dry.

5. Once your homemade indicator papers are dry, cut them into strips for testing acids and bases.

6. Test each liquid from the Tools & Materials list on one indicator paper at a time. See if the indicator turns red for an acid or blue for a base. You can use these indicator strips to test any liquid to find out its chemical property.

HOW AND WHY

The chemical environment inside your body must be kept in neutral balance (meaning not too much acid and not too much base) in order for cells and organs to function properly. The kidneys are in charge of getting rid of acids and holding onto bases to keep levels steady.

Acids are substances with lots of hydrogen ions—particles with an electrical charge—and often taste tangy or sour. Acids will donate a hydrogen ion to other substances. Bases, on the other hand, contain lots of hydroxide ions and are chalky or bitter tasting. Bases will accept or pick up a hydrogen ion from another substance.

While it isn't always possible to taste something to determine if it's an acid or a base, indicator solutions like the one you made in this activity can detect an acid or base by changing colors. Color changes of the indicator solution occur when the acid or base accepts or donates a hydrogen ion. In this chemistry science activity, your cabbage indicator turned pink to red when dipped in acid and blue to greenish-yellow when in contact with a base.

SCIENCE THAT IS SKIN DEEP

The branch of science that studies fingerprints is called dactyloscopy, and it is used to solve mysteries or identify someone who has left a trace print behind. Your fingerprints are formed on the outer layer of skin called the epidermis. Your prints are like no one else's in the world—even identical twins have different fingerprints! For this science and art activity, you'll collect a set of your fingerprints and take a look at the markings that scientists use to figure out whose fingerprint is whose.

TOOLS & MATERIALS

› flour
› clear tape
› sheet of black construction paper
› magnifying glass (optional)

THE STEPS

1. Lightly sprinkle some flour onto a clean surface that your parent says you can work on.

2. Cut 10 pieces of clear tape in equal two-inch pieces, and line them up on the edge of your work surface.

3. Place your black piece of construction paper off to the side.

4. Press each fingertip of one hand into the flour, starting with your thumb and traveling to your pinky finger. Roll your fingertip back and forth a few times to cover your whole fingerprint.

5. Press a piece of tape over each floured finger and then stick it onto your construction paper, starting with your thumb and moving toward your pinky finger. Ask a friend to help you if it is difficult to do this on your own.

6. Repeat with your second hand until you have all 10 fingerprints collected.

7. With a magnifying glass, or by looking closely, see if you can identify these patterns: whorls, loops, and arches.

HOW AND WHY

Fingerprints are small ridges of epidermis (top layer of skin) that form about three months before you're born. These little rows of raised skin cells cause friction that helps you grip and feel things more sensitively. The most common pattern is the loop (65–75% of people), where the lines enter and exit on the same side. Whorls are present in 20–30% of people. Arches, where the lines enter and exit on different sides, are the least common (5–10% of people). Which patterns do you see on your prints?

THE IMMUNE SYSTEM

The organs, tissues, and cells involved in your immune system declare war on invaders that are trying to make you sick. Each day, you share your living space with a community of bacterial colonies, spiky viruses, fungi filled with spores, and squiggly protozoa. Eek! Let's learn about this battle of good versus evil that takes place inside your body and the protective armor your immune system uses to keep out these creepy intruders.

IMMUNE SYSTEM FUNCTION
Defending Your Body

Your immune system's purpose is to prevent you from getting sick by defending against pathogens—microscopic organisms that can cause disease.

Germs can spread through the air, in food or water, on surfaces, or even by hitching a ride with someone or something that can transport them (called a vector), such as a mosquito. Even so, if you think about how infrequently you get sick, you can get a sense of how well your immune system works.

Your body has many layers of protection against microorganisms, beginning with external barriers that make it difficult for bacteria and viruses to enter it. If the enemy (pathogen) breaks past the first safety mechanism and into your blood, there are two more layers of protection that your immune system uses to weaken and remove the unwanted visitor.

Sometimes it takes a few days for your body to win the battle against a virus or other infection. But every time your immune system is matched up against a pathogen, it forms a memory of the trespasser and is even more prepared to fight it the next time. Memory cells help to quickly stamp out an infection that your body has encountered before.

WHOA, WEIRD!

Don't be quick to be grossed out by pus, the yellowish-white thick liquid that oozes out of a cut. Pus is not anything icky. It's a collection of your healthy white blood cells coming to protect the surface of your skin.

FIRST-LINE IMMUNITY
Do Not Enter!

The first barrier that microorganisms encounter is at the entry points in your body: the skin, the nose, the mouth, and the eyes. Have you ever tried to keep your siblings out of your room with Do Not Enter signs, crisscrossing tape, and pillow piles? Your immune system uses the same kind of strategy, called natural barrier immunity, to trap disease-carrying pathogens.

Outer skin layers form a tough, waterproof seal that blocks bacteria and viruses from entering (unless there is a cut or break in the skin). Tears, sweat, and saliva are equipped with chemical enzymes that disrupt bacterial cell membranes. Mucus in the nose and airway form a sticky slime in which organisms get stuck. Cilia, small hairlike attachments in your airway, sweep away pathogens stuck in mucus and help you cough or sneeze them out of your body. Altogether, your barrier immunity functions like a superhero avenger team to keep out diseases!

SECOND-LINE IMMUNITY
Inflammation Station

When a pathogen slips by the first line of defense and enters your blood or respiratory system, the second line of defense springs into action with a strategy called innate immunity. This type of immunity happens no matter what type of bug enters your body; it is a nonspecific immunity. Once your white blood cells sense something foreign in the blood, an inflammation reaction is triggered. An example of inflammation is an itchy rash on your arm. Chemical signals tell your blood vessels to widen and increase blood flow (redness of skin) to the site where the pathogen is. Extra blood flow means extra white blood cells are on their way to help.

After inflammation has been triggered, white blood cells surround and attack the intruders by eating them, a process called phagocytosis. Once the organisms have been swallowed up by the white blood cells, they are digested and chopped into pieces by chemicals called lysosomes. White blood cells also give rise to fever (increased temperature) in the body. By turning up the heat on pathogens, it becomes more difficult for them to function properly and make more of themselves.

THIRD-LINE IMMUNITY
Fighting Cells

To make absolutely sure that your body will recover from an infection, your immune system has a third protection layer to clear germs and keep them away in the future. This type of immunity is called adaptive immunity, and it is targeted to the specific type of microorganism that has sneaked in past the first two lines of defense.

B CELLS: NEUTRALIZING TOXINS

A special type of white blood cell called a B cell identifies the markings on the cell surface of the intruder. B cells then create a custom key-shaped protein called an antibody that floats around in the blood and locks onto the enemy's cell surface marker (antigen), neutralizing the enemy's toxins. Once an antibody is locked onto the enemy's antigen, a signal is sent out to other white blood cells to come and help eat up the intruder.

T CELLS: JUST CALL THEM

Another white blood cell called the T cell is a friend of the B cell. T cells come when B cells signal them. One type of T cell helps destroy infected cells by releasing poison that bursts cell membranes. The other type of T cell calls more white blood cells to come fight the infection. After an infection is under control and the infecting pathogen is dead, some T cells hang around as memory cells. If your body

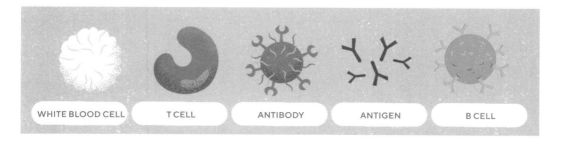

WHITE BLOOD CELL　　T CELL　　ANTIBODY　　ANTIGEN　　B CELL

happens to be exposed to that germ again, memory T cells recognize the enemy and quickly destroy it.

Vaccines (the shots you get at the doctor's office or pharmacy) work by exposing your immune system to a much weaker version of a pathogen so that if you encounter that bacteria or virus in the future, your memory T cells can spring into action and get rid of it before it makes you feel sick.

After passing through three very effective layers of protection by your immune system, most germs don't stand a chance for long in your body.

LYMPH NODES AND SPLEEN
Taking Out the Trash

A clear liquid called lymphatic fluid that travels along a string of pearl-shaped tissues called lymph nodes filters bacteria and other pathogens from the body. It flows through a crescent-shaped organ called the spleen, found in the left side of your abdomen.

Together, the lymphatic fluid and the spleen drain infectious organisms that your immune system has battled and remove damaged or old cells that have died in the process. Your spleen is not a vital organ, meaning you can live without it. However, people who have had their spleen removed are more vulnerable to getting infections with certain types of bacteria.

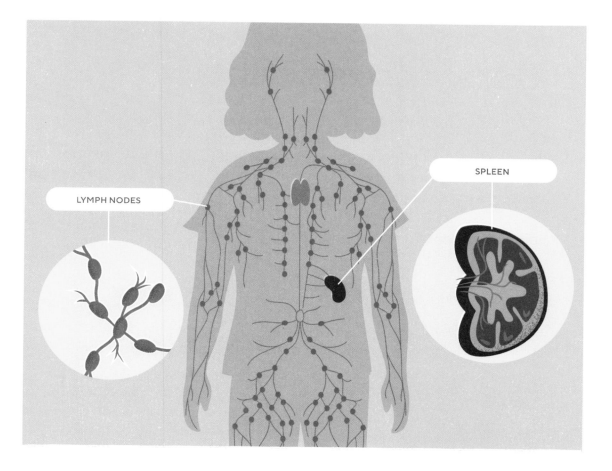

LYMPH NODES

SPLEEN

Putting It All Together

The number one way to take care of yourself and avoid germs is by washing your hands frequently with soap for at least 30 seconds. Face masks work well too; they protect you from respiratory viruses that travel in the air and keep you from infecting others if you are sick. Other important tips to stay healthy: don't share personal items (such as toothbrushes and food), keep up-to-date on your vaccination shots, and make sure food is cooked properly before you eat it. Finally, get at least eight hours of sleep a night to boost your immune system.

ANTIBODY CHALK MURAL

Get out your brightest chalk colors to make a mural of the immune system's cast of characters in this art activity. Bring together the whole crew of viruses, bacteria, white blood cells, and antibodies that interact in the body in a battle of good and evil. Put your own creative storytelling to work with cartoons of how your body's army of white blood cells destroy invading pathogens and clear out infections.

TOOLS & MATERIALS

› chalk in varied colors
› dry outdoor workspace, like your driveway or sidewalk

HOW AND WHY

Your immune system has three layers of defense: B cells make antibodies that neutralize the attack and call T cells to come and help destroy the invader. Macrophages are called to eat and digest the enemy. Lymph nodes and the spleen clear cells that have died in the immunity battle.

THE STEPS

1. Draw and label the different type of white blood cells with different colors in your workspace:

› B cell: a round cell with Y-shaped antibodies stuck to the cell surface
› T cell: a round cell with half-moon-shaped spikes poking out of the surface
› macrophage: a cell shaped like a blob with a mouth for gobbling germ invaders

2. Sketch some viruses and bacteria floating around near the white blood cells.

3. Draw a string-of-pearls chain of white lymph nodes.

4. Draw a purple spleen near the lymph nodes (which is shaped like an orange slice).

5. Add dialogue to your cartoon mural explaining how these cells and organs work together to protect you from germs that are trying to make you sick.

MEET YOUR MICROBIAL NEIGHBORS

One of the major tasks your immune system is in charge of is protecting you from harmful bacteria, fungi, and protozoa (also called microbes) that can cause infection in your body. At the same time, your immune system also learns to live in harmony with "friendly" microorganisms that are beneficial for your body (like the good bacteria that live in your intestines).

Because germs can't be seen without a microscope, it's hard to imagine the trillions of bacteria that live inside of you and on your skin. In this science activity, you will build your own microbiology community and get to know the likes and dislikes of this diverse group of organisms. Watch as colorful colonies multiply, develop, and change into living art that you can see with your naked eye. Graph their growth over time as more species join your block party. This STEAM activity combines microbiology science, living art, and a mathematical graph all in one cool project.

TOOLS & MATERIALS

> enough dirt to fill 4 jars
> shovel
> newspaper sheet or disposable tablecloth for work surface
> large pail or bowl
> 1–2 cups water
> spoon
> 2 raw eggs
> 1–2 cups shredded newspaper or construction paper (no glossy papers)
> 4 glass jars (mason or other) with tops
> clear tape
> black marker
> piece of tinfoil, large enough to wrap around a jar
> 3 steel nails
> toothpaste
> graph paper
> colored pencils

THE STEPS

1. Collect dirt with a shovel.

2. Spread out a disposable cover on your work surface at home.

3. In a large pail or bowl, mix your dirt specimen with ½–1 cup water with a spoon. Your mixture should be loose and slightly damp, not overly wet.

4. Add nutrients—two raw eggs (protein source) and a handful of shredded paper (carbon source)—to the mud. Mix with a spoon or your hands and then wash them well with soap.

5. Spoon the mud mixture evenly among the four jars, leaving one inch at the top of each jar.

6. Label the first jar Sunlight + Air, and place it in a sunny window with the cover on loosely (to let air in).

7. Wrap tinfoil around the second jar and screw the lid on tightly. Label the second jar Dark + No Air, and place it next to the first jar.

8. Add three nails (iron source) to the third jar, and label it Sunny/Air + Iron. Place the cover on loosely, and sit it next to the second jar.

9. With your spoon, make a small well in the mud in the last jar. Squirt some toothpaste (nitrogen source) into the hole. Cover it up with mud. Stand the jar next to the third jar, place the top on loosely, and label it Sunny/Air + Nitrogen.

10. After one day, check your jars to evaluate the water level. Your mud should have settled to the bottom of the jar and water collected at the top of the jar above the mud. You should try to maintain ½–1 inch of water in each jar. Pour off excess water or add a small amount as needed.

11. Watch bacteria in the jars grow over a month. What changes in color and size do you see?

12. Record your observations daily. On graph paper, write the specimen name from your labels as the title of your graph (for example, Sunny/Air + Nitrogen). On the y-axis, plot days 1–30. On the x-axis, plot bacterial growth with colored pencils based on the main bacterial culture color you observe.

13. When you are done with your mud samples, you can add them to an outdoor compost.

HOW AND WHY

Your immune system interacts with an exponential number of bacteria each day, but it's hard to imagine these microbes without seeing them.

You used mud to grow and multiply bacteria and see which organisms your immune system might encounter regularly. You added water and nutrient sources in the form of protein and sulfur from an egg, nitrogen from toothpaste, carbon from shredded paper, and iron from a nail. You developed different conditions for the bacteria to select which colonies are aerobic (prefer oxygen), which are anaerobic (don't like oxygen), which use photosynthesis (sunlight), and which prefer dark conditions. Finally, you recorded your bacterial colony growth over one month using a graph to view their progress.

Each colorful patch that developed on the side of your jar represents a new community of bacteria similar to the ones found on your skin, in your intestines, or in your mouth.

3D MODEL BUILDING

One of the best ways to learn how viruses travel inside your body and cause infection is to build a 3D model of one and look closely at its structure. The COVID-19 virus has an icosahedral (20-sided) shell with spike proteins sticking out of each side that look like a crown (*corona* in Latin) when viewed under a microscope. The coronavirus carries only a few small things (nucleic acids and proteins) inside its shell so that it can copy itself easily once it enters a cell. Using just a few materials, build your own 3D model of a COVID-19 viral particle in this engineering activity.

TOOLS & MATERIALS

› pencil
› 3–4 pieces of cardstock
› traceable triangle template (to the right)
› scissors
› ruler
› clear tape
› piece of string (about 6 inches)
› 3 cotton balls
› red modeling clay
› 20 toothpicks

THE STEPS

1. Trace the triangle template below and cut out the equilateral (same length on each side) triangle.

2. Trace around your triangle on the cardstock twenty times so that you have twenty triangles. Cut them out.

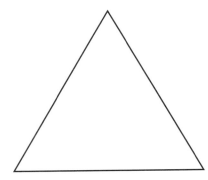

3. Tape five triangles together side to side, moving in a clockwise fashion until you have formed a hexagon (five-sided shape). See the photograph below for guidance.

4. Tape one additional triangle to each side of your hexagon. It should now look like a star.

5. Add one additional triangle with tape at a time, bending and taping the triangle edges together until you have a closed 3D capsule. Right before the last triangle, stuff your string and cotton inside before taping closed.

6. Add a small ball of red clay to the end of each toothpick. These will be the viral protein spikes.

7. Poke a toothpick through the center of each triangle until you have added twenty protein spikes to your COVID-19 model.

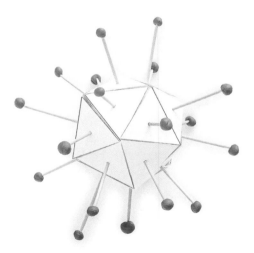

HOW AND WHY

Viruses do not have mito-chondria, a nucleus, and other important structures needed to replicate (make copies) of themselves. They must borrow this type of cellular equipment from you when they enter your body. The COVID-19 virus has a shell with twenty sides called a capsid. Inside the capsid, the virus carries basic infor-mation about itself: nucleic acids (piece of string) and some proteins (cotton balls). Surrounding the capsid shell is an oily (lipid) envelope.

Once inside your body, the virus uses its proteins spikes (toothpicks and clay) to attach to the surface of one of your cells and be drawn inside. From there, it can start using your cell's pro-duction line to manufacture an exponential amount of itself until your immune system starts its war to shut the virus down.

IMMUNITY MATH MINDBENDER

Did you know that you may have *saved someone's life*? It's true! If you have received a vaccination, you have prevented spreading a disease, such as the flu, that could be deadly in someone who is not as healthy as you are. But how do vaccines work to protect large groups of people through immunity? Make two models of immunity and do some simple math calculations to see how vaccines can stop viruses in their tracks.

TOOLS & MATERIALS

› 28 dominoes plus four colored stickers (1 red and 3 yellow), or colored rectangular blocks
› pencil
› sheet of paper
› calculator
› double-sided tape

THE STEPS

MODEL HOW THE FLU VIRUS SPREADS WITHOUT A VACCINE

1. Stand seven dominoes next to each other in one row. In front of that row, stand six dominoes next to each other, then a row of five, and so on until you get to a row with one domino standing up. Your domino arrangement should look like an upside-down triangle.

2. Place a red sticker on the single domino (or if using colored blocks, choose a red one) that is standing up at the tip of the triangle. This will represent the person with the flu. The rest of the dominoes represent people who have not been vaccinated. If using blocks, they should be a color other than red.

3. Knock the red domino over, pointing it toward the others, and see whom the person with the flu infects. Count how many individuals (dominoes knocked over) have fallen ill with the disease.

4. Write No Vaccine on your paper, and calculate how many people (dominoes knocked over) out of 28 total fell ill. What percentage of people became ill? To figure out this percentage, count the number of fallen dominoes (sick people) and divide by the total number of dominoes (all people).

MODEL HOW THE FLU VIRUS SPREADS WITH A VACCINE

1. Set your dominoes up in the same configuration as before, with a red domino at the tip of the triangle.

2. Place a yellow sticker on one domino from the rest of the group. This represents an immunized person who has received a vaccine. Stick a piece of double-sided tape to the bottom of the domino, and put it back in the group (stuck firmly to the surface you are working on).

3. Push the red domino (the person infected with the flu) in the direction of the other dominoes. Count how many dominoes fall. What happened to the dominoes (people) close to the immunized person (the domino stuck to surface)?

4. Write One Vaccine on your paper, and count how many dominoes fell over this time. Divide the number of fallen dominoes by the total number of dominoes (28) to calculate the percentage of people who became ill when one person was vaccinated.

5. Try the same process using three dominoes stuck to the surface with double-sided tape. Knock over the red domino and see how many people become ill when three out of 28 people are vaccinated. Calculate the percentage of fallen dominoes to the total number of dominoes used. What percentage of people became ill now with three people vaccinated?

HOW AND WHY

In this immunity model, you designated one person (the red domino or block) as the individual infected with the flu. Dominoes with double-sided tape that could not fall over (fall sick) represented people with immunity from receiving a vaccine. Vaccinated people don't get sick; they also shield those around them by not passing along the disease. What did you discover when you calculated the percentages of people in your community that are protected with each additional vaccination?

DIY HAND SANITIZER

The best method to keep away germs that can make you sick is by washing your hands for at least 30 seconds with warm water and soap. When you don't have access to a sink, hand sanitizer is a good runner-up. With an adult helping, create a scented hand sanitizer in this science and art activity.

TOOLS & MATERIALS

> ⅓ cup aloe vera gel (may use aloe vera sunburn relief gel)

> medium-sized bowl

> ¼ teaspoon of your choice: peppermint extract, vanilla extract, or almond extract or an essential oil such as lavender

> whisk

> ¼ teaspoon vitamin E oil (optional)

> ⅔ cup 91% isopropyl alcohol

  Note: If you only have 70% rubbing alcohol, change the 2:1 ratio of alcohol to aloe vera to a 9:1 ratio of alcohol to aloe vera

> funnel

> travel-sized squeeze bottles or shampoo bottles or other small containers from your house (empty and washed)

THE STEPS

1. Place the aloe vera gel in a bowl.

2. Choose one of the scents, and whisk it into the bowl. Add the vitamin E oil, if desired, to help soothe dry hands.

3. With adult supervision, pour the alcohol (either 91% isopropyl or 70% rubbing) into the bowl and whisk together with the other ingredients.

4. Using a funnel, pour the hand sanitizer into your containers for use on the go.

HOW AND WHY

Hand sanitizer is a good way to kill germs when you can't wash your hands with soap and water. Not only do your hands come into contact with lots of bacteria and viruses when they touch different surfaces, but you also are likely to touch your face with your hands, spreading the germs to yourself. Cleaning your hands with sanitizer after you touch another person or shared surfaces (like a toilet handle) will help your immune system fight infections.

9

THE ENDOCRINE & REPRODUCTIVE SYSTEMS

The endocrine and reproductive systems release chemical messengers called hormones that travel in your bloodstream and communicate to cells elsewhere in your body. The hormones signal that a change or an action needs to happen. This may not sound too important until you realize that these hormones control your mood, growth, sleep, and lots more. Read on to learn about another example of how your body systems work together as a team to get the job done.

ENDOCRINE AND REPRODUCTIVE SYSTEM FUNCTION
Controlling Change

Organs of the endocrine and reproductive systems are called glands. Unlike the nervous system, where responses happen with lightning speed, the endocrine and reproductive systems control changes that happen slowly and steadily over time in your body. Gland control centers are located in different areas of your body, including the brain, the neck, the abdomen, and the groin.

Chemical signals are released into the blood as hormone molecules. These traveling news reporters head out to various areas in the body and tell targeted cells to make specific changes in their behavior or production—either up or down a notch.

When hormones reach their destination, they fit and lock onto a special site called a receptor on the type of cell they want to communicate with. After connecting to a receptor, the hormone can give feedback to the cell about either increasing or decreasing the work it is doing. In this way, hormones can control your mood, sleep, growth, and development of your body into an adult, and how your body is using oxygen and sugars for energy.

There are lots of checks and balances that the glands use to decide how much hormone to release, including the presence of stress or infection in your body. Glands also sample the blood to get a sense of how much minerals, fluid, and circulating hormone is already present and moving around.

In addition to producing important chemical messengers, the reproductive system and its organs allow a person to pass their genetic information with all of their traits, such as height and hair color, to their child. Although there are about 30 trillion cells in your body, all of the genetic information you inherited came from just two cells made by the female and male reproductive glands.

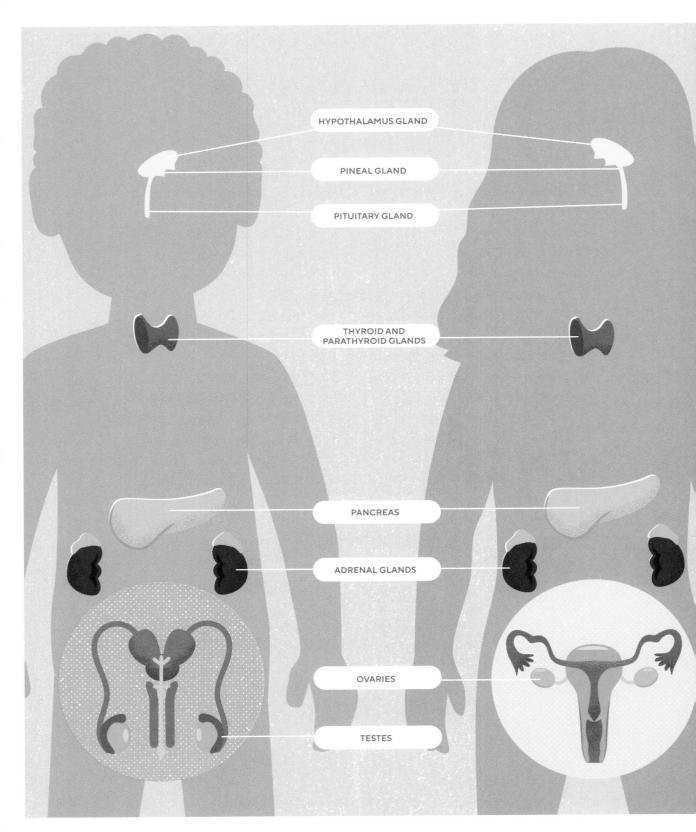

HYPOTHALAMUS GLAND

PINEAL GLAND

PITUITARY GLAND

THYROID AND
PARATHYROID GLANDS

PANCREAS

ADRENAL GLANDS

OVARIES

TESTES

HYPOTHALAMUS
The Hormones' Boss

The hypothalamus is a structure in the lower brain that links the nervous system to the endocrine system. Much like other parts of the nervous system, the hypothalamus receives information from your environment—if it is sunny or dark, hot or cold—and makes decisions about how to respond. The hypothalamus releases hormones that control thirst, hunger, body temperature, and fluid balance at sites throughout your body.

The hypothalamus also supervises one of the main endocrine control centers: the pituitary gland. Depending on what actions or changes are needed to keep a steady, balanced environment for your cells to do their work, the hypothalamus tells the pituitary to stop producing or make more of its hormones.

PITUITARY GLAND
Little Gland, Big Job

Even though it's as small as a thumbtack, the pituitary gland is the big cheese of the endocrine system. It sends out many different types of hormones into the bloodstream that act on other glands throughout the body. From its seat in your skull (behind the nasal cavity), the pituitary directs the adrenals, thyroid, ovaries, testes, and mammary glands to speed up or slow down the work they are doing. The pituitary also sends out a growth hormone, which encourages your muscles and bones to develop into adult size, and a fluid-regulating hormone (ADH) that encourages your kidneys to hold on to water.

WHOA, WEIRD!

There can be too much of a good thing. Although rare, when the pituitary produces too much growth hormone, it can result in gigantism—growing too tall.

Hugs, snuggling with pets, and exercise trigger the pituitary to release endorphins, a hormone that lessens your ability to feel pain. All the more reason to show some love!

THYROID AND PARATHYROID GLANDS
Keeping an Eye on Energy

Just below your Adam's apple sits your thyroid and parathyroid glands. Shaped like a butterfly, the thyroid wraps around your windpipe (trachea) and has four parathyroid glands tucked into its corners.

The thyroid controls how energy is used by your cells (called metabolism). A delicate balance of thyroid hormone is very important: too much can cause jitteriness, fast heart rate, and weight loss, while too little can result in weight gain or leave you feeling overly tired and cold. The pituitary gland keeps a close eye on the thyroid to ensure just the right amount of thyroid hormone in the blood. The parathyroid glands keep calcium levels stable in your blood to nourish your bones and muscles.

WHOA, WEIRD!

Teenagers are known for big mood swings and eating all the snacks. This is a normal response to a big surge and shift in the level of hormones during puberty that control appetite and feelings.

Have you ever woken up exactly on time without an alarm clock? Melatonin, released by the pineal gland, acts as an internal timer, letting your body know the time of day and season without outside cues.

Is there a peanut-free zone at your school? Some people have a dangerous allergy to peanuts that can result in an emergency with their breathing or blood pressure. An epinephrine (epi-) pen can be used to give a large dose of adrenal gland hormone that can jump-start a rescue process in the body and save the person's life.

ADRENAL GLANDS
Guarding Against Danger

You know that feeling you get when you lift a rock up and find a small snake? The startle response that makes your heart race and gets you ready to sprint away from danger is your adrenal glands in action. This protective mechanism is called the fight-or-flight response, and it readies your body to escape quickly from danger. The fight-or-flight response is jump-started when your adrenal glands release a hormone called adrenaline.

Adrenaline activates the sympathetic nervous system and gives you a burst of energy in case you need to get out of a situation quickly (or think you do). Steroids are also released by the adrenal glands. These hormones act in many locations in the body to help your body respond to salt imbalances in the blood, stress, and infections, and grow into adult size.

PINEAL GLAND
A Hidden Gem

The pineal gland is an understated but important little gland that hides deep in the middle of your brain. Melatonin is secreted (sent out to the bloodstream) from the pineal gland. This hormone is responsible for your sleep–wake cycle and supports a regular menstrual cycle in females.

PANCREAS
Turning Sugar into Energy

You have met the pancreas before in the digestive system chapter. This gland, located next to the stomach, releases insulin and glucagon into your bloodstream. Sugars (in the form of carbohydrates) that you eat are processed by these two hormones and converted into active energy for your cells or stored for later use.

The pancreas can be overwhelmed by a diet that contains too much sugar. Excess sugar levels in the blood over many years can shut down the pancreas from making insulin and poison the cells in your nervous, excretory, and cardiovascular systems. The result is a disease called diabetes.

REPRODUCTIVE GLANDS
Growing Up Glands

Together, the pituitary and reproductive glands change the look of your body as you grow older and go through puberty. The reproductive glands are different in males and females. The testes, found in the groin in males, make the hormone testosterone. Testosterone helps to develop the body characteristics of a boy into those of an adult male. It also helps to produce sperm, the cells that carry a father's genetic material (a map of his traits).

The ovaries, found in the lower abdomen in females, are the female reproductive glands that produce progesterone and estrogen. These two hormones help develop the body characteristics of a girl into those of an adult female and control the menstrual cycle. Once a month, progesterone and estrogen trigger an egg cell to develop in the ovary; this egg carries the mother's genetic material that can combine with a sperm cell to produce a new human.

Putting It All Together

Taking care of your endocrine and reproductive systems comes down to a few simple, healthy habits. In order to maintain the delicate hormone balance of your endocrine system, make sure to get enough sleep, drink enough water, and take quiet time off screens to de-stress. Eat foods that are not too high in processed sugars (check the "added sugars" label on the back of the food) to be kind to your pancreas and avoid diabetes.

Annual checkups with your doctor will make certain that all the endocrine and reproductive glands are functioning properly. Take care of your reproductive system by asking a parent or other trusted adult honest questions about changes happening in your body. This is your best strategy to become informed about ways to stay healthy as a teenager.

RUBE GOLDBERG ENDOCRINE MACHINE

The endocrine system communicates through a series of chain reactions in your body, with one gland talking to the next like a game of telephone. In this engineering activity, let's build a Rube Goldberg machine, a chain-reaction-type contraption designed to perform a simple task in a goofy, complicated way. Starting with the hypothalamus, build a machine that follows hormone release from gland to gland until the final action at its targeted receptor site.

TOOLS & MATERIALS

YOUR CHOICE OF

›   clean materials from the recycle bin: foil, bottles, containers, cardboard, food boxes, cans, toilet paper or paper towel tubes, ribbons, fabrics, tissue boxes

›   materials that can move: marbles, dominoes, small balls, toy cars, string, levers, magnets, inflated balloons (that can deflate), old CDs

›   other materials to consider: books, feathers, buckets, bowls, tape, toy train or car tracks, building blocks, Legos, fans, zip ties, action figures

ALSO GATHER

›   pencil
›   sheet of paper
›   black marker
›   tape for labels
›   phone, tablet, or other video camera source (optional)

THE STEPS

1.   Sketch out on a sheet of paper the following hormone path from gland to receptor target:

HYPOTHALAMUS	→	PITUITARY GLAND	→	THYROID	→	HEART CELLS	→	HEART RATE INCREASED!
Hormone		Hormone		Hormone		Receptor site		Action!

2. Sketch out five steps in your Rube Goldberg machine on paper to demonstrate this chain reaction.

3. Build your machine (a path through tubes and containers) out of your selected materials. Place it on the floor.

4. Label each point in the chain reaction with tape and a black marker on the floor next to your machine so the travel path of the hormone to the receptor target can be tracked.

5. Send one of your materials that moves through your machine.

6. Watch (and record, if you want) the chain reaction! It may take several tries and several video clips to catch a complete Rube Goldberg reaction.

HOW AND WHY

Starting with the hypothalamus in the brain, you followed thyrotropin-releasing hormone to the pituitary, where thyroid-stimulating hormone was released to make the thyroid secrete thyroxine hormone, which travels to the heart cells and has an action of increasing the heart rate! Each endocrine chain reaction was modeled by your Rube Goldberg machine actions. How many times did it take you to get your final video clip of the whole sequence?

FIGHT-OR-FLIGHT

Do you hear a dog barking loudly that you don't recognize? Is it almost your turn to speak in front of the entire school? Did you uncover a squirming cluster of maggots under a rock? Think about how your body would react in these situations. The reflex that turns up your heart rate, gets your leg muscles ready to run, and gives you a jumpstart of jittery energy is called the fight-or-flight response. It happens when the adrenaline hormone is released into your bloodstream. Let's test this response with a creepy physiology (science) experiment.

TOOLS & MATERIALS

› stopwatch or phone timer
› pencil
› sheet of paper
› computer with video link

THE STEPS

1. Sit in a quiet place and use your right hand to find the pulse (heartbeat) in your left wrist. Feel for your pulse at your inner wrist near your thumb side. You may have to press a bit to feel it. Set your stopwatch for 30 seconds and count how many beats you feel. Multiply this number by two to calculate number of beats in one minute.

2. Ask a family member or friend to hide in your home and jump out at you and scare you. Walk from room to room with a stopwatch until you encounter the startle. Stop where you are and sit down to feel and count your pulse again for 30 seconds. Multiply this number by two to calculate beats per minute.

3. Play the trust fall game. Have a partner stand a few feet behind you. Stand with your feet together, legs stiff, and arms folded across your chest, facing away from your partner. Your partner stands in a lunge position facing your back, with both arms slightly bent in front of their chest and palms facing upward. When your partner says they are ready, fall backwards into their arms. When it's over, time your pulse again for 30 seconds, and calculate beats per minute one last time. How do all your heart rates compare?

HOW AND WHY

You likely felt scared for a minute during the snake video and when you were startled in your home. What did you notice about your heart rate then versus when you were sitting quietly? Did you notice any other body sensations? What were they? When the fight-or-flight response is activated, your pituitary gland stimulates your adrenal glands to release a hormone called adrenaline into your bloodstream. Adrenaline works at many receptor sites in the body to increase heart rate, increase blood flow to muscles, and give you a burst of energy to help you escape in a true emergency.

HORMONE-RECEPTOR MATCHING GAME

Each hormone in the endocrine system has a lock-and-key fit to a receptor at its target cell. After the hormone–receptor match is made, an action takes place in the body. In this science-based game, flip over two cards at a time, looking for a match, and learn where hormones of the endocrine system pair to receptors.

TOOLS & MATERIALS

› **24 equal-sized, blank paper square cards**
› **scissors**
› **ruler**
› **colored pencils**
› **black pen**

THE STEPS

1. Make 24 equal-sized square cards. Take pairs of cards and shade them the same color with your colored pencils to make twelve pairs. Place an H for hormone at the bottom of one colored card and an R for receptor at the bottom of the other card of the same color.

2. Write the names of the following hormone and receptor pairs to make a match of the same-colored cards.

› Thyrotropin-releasing hormone (H) and Pituitary (R)
› Thyroid-stimulating hormone (H) and Thyroid (R)
› Melatonin (H) and Skin (R)
› Oxytocin (H) and Uterus (R)
› Estrogen (H) and Ovary (R)
› Testosterone (H) and Muscle (R)
› Adrenaline (H) and Heart (R)
› Cortisol (H) and Fat tissues (R)
› Calcitonin (H) and Bone (R)
› Insulin (H) and Pancreas (R)
› Growth hormone (H) and Liver (R)
› Prolactin (H) and Breast (R)

3. Turn over the cards so the blank sides are showing, and mix them up. Spread them out in a random order or in four rows of six, whichever you prefer.

4. Recruit a partner to play your endocrinology matching game and look for matches. Each time you find two same-colored cards that connect a hormone to its corresponding receptor site, you win that pair. Whoever wins the most pairs wins the game.

HOW AND WHY

Hormones control many actions in the cells of your body. They are produced in endocrine system organs called glands and travel to the site of their targeted action. Once a hormone links to a receptor, a chemical reaction takes place, doing the job the hormone set out to do. An example is when calcitonin is released by the thyroid gland and travels to bone cell receptors to decrease the amount of calcium in the blood.

DNA EXTRACTION EXPERIMENT

What makes you uniquely you? The secret code to all your inherited traits is carried inside your cells on structures called genes. Genes are made of a chemical called DNA that is shaped like a beautiful spiral staircase called a double helix. It isn't just humans who have traits passed down on DNA; fruits, vegetables, and plants do as well. In this kitchen-lab science activity, you will learn how to extract DNA from strawberries.

TOOLS & MATERIALS

› ½ cup frozen strawberries
› 2 tablespoons pineapple juice or meat tenderizer powder
› resealable plastic bag
› rolling pin
› tall clear glass or clear test tube
› ⅔ cup rubbing alcohol (70%)
› wooden skewer

Caution: Do not taste or drink any of the mixture; the alcohol is toxic.

THE STEPS

1. Put the frozen strawberries and either the pineapple juice or meat tenderizer powder in a resealable plastic bag. Mix well.

2. Crush the frozen strawberries with a rolling pin or with your hands on the outside of the bag. The liquid should be soupy and thick—not too thin or overmixed.

3. Pour the liquid into a clear glass about halfway up.

4. Gently and slowly pour the rubbing alcohol into the glass, taking care not to mix the alcohol in with the fruit mixture.

5. Check back in about 10 minutes and observe a white film at the top of the alcohol. This is your strawberry DNA! It will look like squiggly white fibrous strands. Use the wood skewer to retrieve it from the glass to observe.

HOW AND WHY

Freezing fruit disrupts the cell wall of the strawberry cells. DNA is released into the juice after crushing the fruit, and an enzymatic chemical reaction from the pineapple juice/meat tenderizer helps to uncoil the DNA. Adding alcohol to the liquid separates the DNA from the juice mixture with a process called liquid extraction.

Farmers and agricultural scientists are interested in fruit and vegetable traits, just as geneticists are interested in human traits. Some strawberries are tart and some are sweet, some are firm and round while others are softer and bruise easily. Traits that farmers are interested in keeping can be selected for by studying the DNA after extraction, like the extraction you just completed. Likewise, biochemical scientists are studying gene therapy in humans to fight diseases by making changes to DNA inside of cells. Cool!

WE ARE FAMILY!

Has anyone ever told you that you have your mom's nose or your dad's height? Our physical traits (the way we look) are passed down to us (inherited) from our parents and grandparents through cell structures called genes. Let's find out from whom you got your good looks by tracing your heredity in this art-related activity.

TOOLS & MATERIALS

› family photo album
› colored pencils or pens
› 2 sheets of paper

THE STEPS

1. Work with a family member to find pictures of your mother's parents and your father's parents.

2. Record notes about eye color, hair color (or baldness), hair texture (fine, thick, straight, wavy), skin tone, nose shape, height, and face shape of your four grandparents, your mother and father, and your siblings.

3. Draw your family tree using the template on the next page. Add details to your drawing from your notes above. Can you tell from whom you and your siblings inherited the above traits?

HOW AND WHY

Although there are trillions of cells in your body, all of your inherited information comes from two cells only, the sperm from your biological father and the egg from your biological mother. Those two cells combine to form a fertilized cell that has 23 chromosomes from your dad and 23 chromosomes from your mom. Chromosomes are chemical structures that contain genes—the map of all your inherited information. Depending on how the genes from your parents (and their parents) mix, certain features (like freckles!) will show up in you.

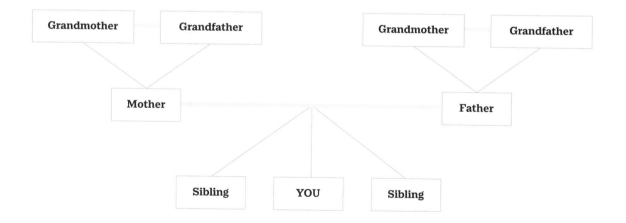

Conclusion

YOUR BODY AND YOU

Now that you have seen up close the incredible science, technology/design, engineering, artistic beauty, and mathematical accuracy that is required to run your body systems, take a minute to appreciate yourself and the phenomenal things your body can do. There is only one person like you in the entire world, and you are amazing!

RESOURCES

BOOKS

First Human Body Encyclopedia
This excellent reference book has interesting images for kids.

Joe Kaufman's How We Are Born, Grow, Work and Learn
A classic children's resource describes the human body systems with interesting facts and engaging cartoon illustrations.

Illumanatomy by Kate Davies
See inside the human body with a magic viewing lens. Enjoy a visual tour of the body systems at the different tissue and organ levels.

Netter's Anatomy Coloring Book by John T. Hansen
This book uses a hands-on, art-focused STEAM approach to learning about the human body systems, tissues, organs, and cells.

ONLINE

The Breakfasteur
www.youtube.com/channel/UC7J5njZ1UHnmKLuNSToKkUQ
Supercool playdough surgeries with Dr. So and her children give an up close view of the body systems.

Ck12.org
www.ck12.org/student
Visit for well-organized human biology sections about the body systems for kids ages 8-12.

Operation Ouch
https://goo.gl/iDvxKs
View this fun science show for kids with experiments by Dr. Chris and Dr. Xand, highlighting the human body systems.

INDEX

Page numbers followed by "f" indicate figures; page numbers followed by "t" indicate tables.

ACKNOWLEDGMENTS

Find yourself a partner who gets you; who does not try to dissuade you from big ideas, dreams, and biting off more than you can chew; whose value set aligns with yours; and who has perfected the best morning coffee and hug as you leave for work. Jeff, you are all of those things and more. Your support means everything.

Jon and Adam, well, I guess it didn't turn out too badly for me in the sibling lottery. I suppose it is fair to let you know now how much I look up to both of you and that you were truly the inspiration for this book. When we were kids, I watched as you both built by hand, puzzled your way, invented, or debated Mom and Dad to create all the fun we had without screens and technology. From go-carts, magic tricks, twelve-foot forts, stilts, and beyond, you taught your little sister that the sky was the limit for what you could achieve or adventure into, even as a young person. As an adult, you convinced me that with grit, curiosity, and perseverance, I could succeed at anything—even completing medical school and two residencies while raising two small children. Without you behind me, I most certainly would not be right here at this wonderful spot in life right now.

Julia, Tanya, Sarah, and Shannon, when I am stuck in a loop of uncertainty or blocked from heading forward, all I have to do is lift my head up a bit and turn my gaze toward the four of you, and I am inspired and resilient again. "Movement creates movement," "If it's meant to be, it's up to me," "Where's the silver lining?" and "Bring on all the crazy, I love it all," you tell me. Each of you tackles professional goals, family balance, and personal growth in ways that astound me. If ever I am in a pickle, I listen in my head to the pieces of advice you have given me over the years, and I know how to weather the storm. You are my biggest fans and the steady ground beneath my feet. I am deeply indebted (and the better person) for your friendship.

Elle: Spending time with you as your mentor has made all the difference in how I view our professional roles in medicine—the science of caring for others. Our friendship has pushed me to think beyond just a career title and more about what I *want* to do and am *inspired* to do, including writing this book. The future has more vibrancy and glimmer with you in it.

Remembering my parents, Alice and Thomas Cheyer, who were nerdy, counterculture cool before it was cool and who taught us that books are invaluable treasures. Also acknowledging the entire Cheyer/LaFleur extended family for supporting my crazy plans over the years.

Meg Ilasco, two Pisces fish swimming freely represent the highest lesson of fellowship and the capacity to see deeply into the life and nature of things, teaching us not to be afraid to let go of our earthly form and dream. From the very first time we worked together, I knew we were connected in our thinking and trusted each other implicitly. Thank you for believing in me at the start and for working with me on all future creative endeavors, which I plan to hold you to!

Susan Randol, your depth of experience and passion for your work as an editor fueled this project and kept the train rolling forward. You were kind yet laser sharp, forgiving yet meticulous. It was such a pleasure to work with you on this book. It is clear that one of your superpowers is knowing how to bring out the best in an author. I am very grateful for your investment in this manuscript.

From the polished, creative team at Penguin Random House, I offer my most heartfelt appreciation for designer Katy Brown, whose keen eye pulled together a beautiful story image on the pages of this book; illustrator Diego Vaisberg, whose artwork speaks a language that kids naturally understand; photographer Nancy Cho, who brought the STEAM activities to life; and copy editor Jane Katirgis.

For encouraging me to move outside my comfort zone as a physician and pursue other professional aspirations, I thank The Launch Collective Sonoma County. The resources, pearls, networking, and community you have provided have been invaluable.

With the memory of Micah Price front of mind, I acknowledge the incredibly impactful work of the teachers at The Healdsburg School and the teachers across the United States who are devoted to instilling a love of learning in our children.

About the Author

Dr. Sara LaFleur is a physician in the San Francisco Bay Area who cares for both children and adults. She earned her MD from Tufts University School of Medicine and her BA in psychology from Emory University. Sara is an advocate for young adults looking to enter STEM careers and is a professional mentor for the American Medical Women's Association. She has won several teaching awards for her commitment and excellence in training students at a Harvard Medical School teaching hospital. She is a founding member of the Launch Collective Sonoma County, a community of female entrepreneurs and business owners. Sara enjoys spending her free time with her husband, two daughters, and dogs hiking along the east and west coasts and seeking out the best local fresh fruit pies. Sara thrives on connecting with others and living her life with determination, empathy, and a drive for adventure.